# Boreal Ecology

# Three
# week loan

Please return on or before the last
date stamped below.
Charges are made for late return.

© William O. Pruitt, Jr.

*First published 1978*
by Edward Arnold (Publishers) Limited
25 Hill Street, London W1X 8LL

Board edition ISBN: 0 7131 2685 X
Paper edition ISBN: 0 7131 2686 8

Printed in Great Britain by
The Camelot Press Ltd, Southampton

# General Preface to the Series

It is no longer possible for one textbook to cover the whole field of Biology and to remain sufficiently up to date. At the same time teachers and students at school, college or university need to keep abreast of recent trends and know where the most significant developments are taking place.

To meet the need for this progressive approach the Institute of Biology has for some years sponsored this series of booklets dealing with subjects specially selected by a panel of editors. The enthusiastic acceptance of the series by teachers and students at school, college and university shows the usefulness of the books in providing a clear and up-to-date coverage of topics, particularly in areas of research and changing views.

Among features of the series are the attention given to methods, the inclusion of a selected list of books for further reading and, wherever possible, suggestions for practical work.

Readers' comments will be welcomed by the author or the Education Officer of the Institute.

1978

The Institute of Biology,
41 Queens Gate,
London, SW7 5HU

# Preface

As the world's human population continues its dizzy upward spiral, man increasingly invades ecological associations that are farther removed from the so-called human optimum, the temperate zones. Just as voles and lemmings are restricted to optimum habitat areas when their populations are low, but spill over into marginal habitats when their populations are high, so the human population is now spilling over into more northern regions. Temperate-zone man is encountering conditions new to his experience. Thus Boreal Ecology is a timely subject.

*Boreal* means simply 'northern' and is a relative term used in distinction to *austral* or southern. Perhaps a good ecological definition of the boreal regions is: those where snow cover affects animals and plants or where living organisms have evolved adaptations to snow.

Because of the frequent juxtaposition of 'boreal' and 'coniferous' when describing the taiga one should be alert not to restrict the meaning of 'boreal' to association with coniferous forest vegetation. Remember that the tundra is more 'boreal' than is the 'boreal forest'.

Many of the ideas presented here were first encountered or were honed by visiting lecturers to my course in Boreal Ecology over the past fifteen years, or by the students themselves in discussions or laboratory and field projects. My wife Erna has given many valuable ideas and critical comments. Shirley Lowry transcribed tapes of classroom discussions and typed the manuscript. Wolf Heck made the drawings and photographs. To all, I am grateful.

Winnipeg, Manitoba, 1978

W. O. P.

# Contents

# 1 Tundra, Taiga and Other Northern Words

## 1.1 Arctic

Most maps give a distorted view of Canada and other northern regions. The North Polar Projection is the only type of map that gives a correct impression of the Arctic Basin and the lands surrounding it (Fig. 1–1).

Note in Fig. 1–2 the vast difference in amount and seasonal variation of incoming energy according to latitude. Note in particular the four closest curves, which, for part of the year, drop below the zero line. Between the third and fourth curve is a region where for one day in winter the sun does

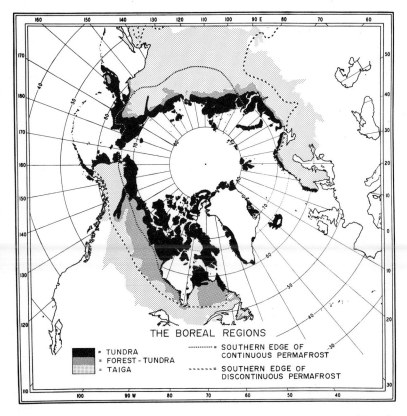

THE BOREAL REGIONS

■ = TUNDRA
▨ = FOREST - TUNDRA
░ = TAIGA

·········· = SOUTHERN EDGE OF CONTINUOUS PERMAFROST
------ = SOUTHERN EDGE OF DISCONTINUOUS PERMAFROST

**Fig. 1–1** North Polar Projection, emphasizing the Boreal Regions. At this scale boundaries are approximate.

not rise, and for one day in summer it does not set. This is the Arctic Circle, 66° 30′ North Latitude. To a geophysicist or astronomer the *Arctic* is the region north of the Arctic Circle. This is a rather arbitrary and artificial delimitation.

The climatologists' definition of Arctic is a little better for our purposes, but still not entirely satisfactory—the region with a mean

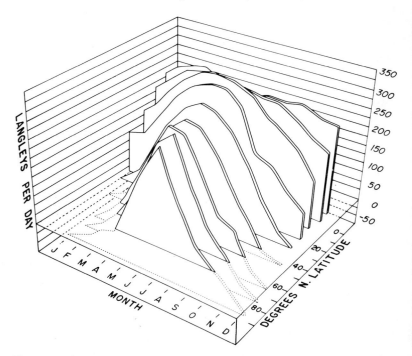

**Fig. 1–2** Three-dimensional representation of energy flux at the earth's surface at various latitudes throughout the year. Points to note: massive decrease in incident energy in boreal regions compared with temperate and tropical regions; marked annual periodicity in boreal regions; negative energy balance during part of the year in boreal regions. (Data from BUDYKO, 1963.) (1 Langley = 701 Wm⁻²)

temperature of no more than 10°C for the warmest month. When one plots on a map the southern border of this region there is a reasonably close agreement with the polar tree line.

So boreal ecologists are reduced to devising their own definition of Arctic. Most biologists use the term Arctic to include all the region lying north of the poleward limit of trees. There are some important exceptions as shall be seen later. Moreover, the definition is suitable only for the land. In the sea there is an entirely different set of relationships (DUNBAR, 1947a).

## 1.2   Subarctic

*Subarctic* is a rather nebulous term with a variety of definitions. The climatologists' definition of subarctic is that region in which the mean temperature exceeds 10°C for not more than four months of the year and where the mean of the coldest month is not more than 0°C. Again, this definition results in some important anomalies.

Another definition looks on the subarctic as the region with both forest and continuous or discontinuous permafrost. This is a very reasonable definition and comes closest to the concept most English-speaking biologists have of the subarctic. This kind of forest is usually quite open, of almost park-like aspect. For examples of the complexity of the terminology see BLÜTHGEN (1970) and HARE and RITCHIE (1972).

## 1.3   Tundra

Another common northern word is *tundra* (pronounced TOON-dra). It originated from the Finnish *tunturi*, the plateau in northern Finland, without trees and dissected by deep, forested valleys. Tundra simply means a land without trees or the land beyond the poleward limit of trees, and actually consists of many vegetation types.

## 1.4   Low Arctic and High Arctic

Two confusing terms are *Low Arctic* and *High Arctic*. These refer to latitude, not altitude. Low Arctic (Fig. 1–3) is the more southern part of the tundra, with complete vegetation cover, occasional patches of tall shrubs, much snowfall and much wind. There may even be a winter rain or a short thaw. High Arctic (Fig. 1–4), on the other hand, is the more northern tundra with no vegetation taller than about shoe-high, much bare ground, little snow, little wind and no winter thaw. This is the true 'Polar Desert' and is actually quite dry (FREUCHEN and SALOMONSEN, 1958).

It must be noted here an important difference between the Russian and English language use of these terms. In the Russian ecological literature subarctic and Arctic refer to what are called Low Arctic and High Arctic, respectively, in English (GRIGOR'EV, 1946). It is important to recognize the difference because of the many valuable Russian publications in Boreal Ecology and because of the great advances made in the USSR in understanding their north country.

## 1.5   Taiga

*Taiga* (pronounced TIE-gah) in Russian originally meant a dense marshy forest in Siberia. At one time some North American botanists tried to restrict the meaning of taiga to the open park-like forest just south of tree line but this restriction has been rejected by all boreal ecologists

**Fig. 1–3** Low Arctic Tundra. Points to note: the haze in middle distance is siqoq; in left foreground is a zaboi filling a creek bed to thicknesses of 5 to 8 m; in right foreground the mottled area consists of cotton sedge (*Eriophorum*) tussocks; the dark area in middle distance is a vyduv or an area continually blown clear of snow. Ogotoruk Valley, north-western Alaska.

(HOFFMANN, 1958). The present meaning of taiga is the boreal or northern coniferous forest (Figs. 5–1 and 5–2), essentially the original Russian meaning, but extended to other regions in the northern hemisphere (BRYSON, 1966; BALANTSEVA, 1960; ZHUKOV, 1966).

## 1.6 Forest-tundra

*Forest-tundra* is the ecotone between taiga and tundra. Although extensive in North America it is not very important economically and consequently has been little studied. In northern Eurasia, especially in USSR, the forest-tundra is extensive (BALANTSEVA, 1960) and important since it forms the main winter range of reindeer (AHTI, 1961). The Soviets look on the forest-tundra as a separate ecological zone rather than just an ecotone. There are three types of forest-tundra: isolated clumps of forest surrounded by tundra (Fig. 1–5); scattered trees interspersed with tall shrubs (Fig. 1–6); extensive stands of willow and birch shrubs (Fig. 1–7).

In eastern North America and north-eastern Asia there is another type of forest-tundra: areas covered with dense, stunted shrub-like trees,

**Fig. 1–4** High Arctic Tundra. The flat area in middle foreground is primarily sedge (*Carex*) meadow, traversed by low beach ridges; the high area beyond the river gorge, as well as the distant plateau, are Polar Desert. The haze in middle distance is siqoq. Devon Island, NWT.

pruned and shaped by the wind. This is called *tuckamoor* or 'tuck' and its ecology is very poorly known.

## 1.7 Conclusion

One important aspect of the ecology of northern regions is the fact that these regions are young; they only recently have been exposed by the retreat of the Pleistocene ice sheets (CRAIG and FYLES, 1960). Indeed, some areas, such as Greenland and parts of the Canadian Eastern Arctic, are still in this condition. Man lives now in the Pleistocene, in an interglacial, and all plants and animals in northern regions are still affected by glaciation. Thus, their ranges are shifting and expanding; they are coming into contact with species never before encountered (BRYSON, IRVING and LARSEN, 1965; BRYSON and WENDLAND, 1967).

During earlier glacier expansions there were different distributions of vegetational zoning than at present. The detailed evidence is conflicting and there is no agreement. It is known that in earlier times there were environments that are now extinct. For example, in parts of Eurasia there were cold, dry steppes that graded into sedge-lichen tundra to the north

**Fig. 1–5**  Forest-tundra. Grove of *Picea mariana* in protected site. Note the layering and flagging of the black spruce in foreground, caused by siqoq; note caribou trails in sedge meadow at foot of ridge. East side of Artillery Lake, NWT, at Last Woods, the place where Ernest Thompson Seton first saw 'The Arctic Prairies'.

through a kind of forest-tundra. This cold steppe is extinct today, along with the animals that used it. Many species and associations of species perhaps evolved in different combinations than they occur today. Thus the boreal biotic associations are still young. Animal and plant communities are perhaps unsaturated, so that different combinations of animals and plants should be expected.

Boreal ecologists deal with aspects of nature, particularly snow and ice phenomena, for which there are no precise English words. Consequently our writings and speech are larded with Inuit, Athapaskan, Lappish and Tungus words, not in any attempt to be erudite but to aid precision in our speech and thoughts.

**Fig. 1–6**  Forest-tundra. Scattered black spruce with understory of willows. Old John Lake, north-eastern Alaska.

**Fig. 1–7**  Forest-tundra. Extensive stands of willow (*Salix* spp.) and dwarf birch (*Betula glandulosa*). Such areas are important as rutting grounds for caribou. Southern slopes of Baird Mountains, north-western Alaska.

# 2 Radiant Energy and Light

## 2.1 Radiant energy

In boreal regions one is acutely aware that the statement 'The sun is the source of all life' is no idle cliché but a basic ecological fact. The solar constant, 1360 Wm$^{-2}$, is the energy that reaches a plane surface at right angles to the sun's rays at the top of the earth's atmosphere. Various wavelengths are filtered out and the total amount of energy reaching the actual surface of the earth is reduced by reflection and refraction from water and dust. The energy that finally reaches northern regions must traverse a longer column of air than does the energy that reaches tropical regions. Moreover, the general angle of incidence is much lower in northern regions than in equatorial and temperate regions. When the sun is 90° above the horizon 92% of the energy received at the surface is direct radiation, but when the sun is near the horizon (8°) 50% of the energy received is by diffuse radiation (ALLEE, EMERSON, PARK, PARK and SCHMIDT, 1949). Thus the energy that finally reaches the earth's surface in boreal regions is considerably less than that which reaches tropical surfaces.

Because the earth's axis tips, the boreal regions are turned first towards and then away from the sun. The result is a violent pulsation in available incoming energy from some 300–500 langleys/day in mid-summer to zero (or even less) in mid-winter (HARE and RITCHIE, 1972). During the period of winter darkness north of the Arctic Circle there is actually a loss of energy to space (Fig. 1–2). This is more than made up for by the incoming radiant energy during the remainder of the year so that no part of the boreal regions has a total negative radiation balance (ORVIG, 1970; DOLGIN, 1970).

Of the wavelengths that penetrate the atmosphere the infrared portion contains most of the heat energy in boreal regions, but in autumn, winter and spring, when the sun may be below the horizon for long periods, the heat energy flows back out to the infinite heat sink of space. If the sky is clear and the air dry, there is little to prevent radiant heat loss to space; if a thin skiff of cloud moves over the site then the infrared is reflected back to earth. Thus the air temperature may remain the same (say, −40°C) but the effective radiation temperature may vary from −40° to −90°. Consequently, the readings of the traditional thermometer are, in boreal ecology, not nearly as important as the readings of a radiometer (STOLL, 1954; HARDY and STOLL, 1954).

Once one learns to think in terms, not of degrees on a thermometer, but of joules of heat flowing from a source to a receiver or accumulating body, then many aspects of boreal ecology are easier to understand. Heat flows by convection, conduction and radiation. In temperate and tropical regions convection and conduction are the important routes of heat flow but as one goes farther north radiant energy exchange becomes more and more important. When considering organisms and their environment one must evaluate this flow of energy (GATES, 1962). The organism's response may have little to do with the readings of an alcohol thermometer. Most measurements of basic climate and weather parameters are not directly usable by boreal ecologists, but must be related to such phenomena as moisture transfer, heat transfer, carbon dioxide, etc.

For example, a widely-held piece of folk-lore is that many northern animals are white in order to conserve heat. But studies have shown that all boreal mammals and birds tested are 'black bodies' as far as infrared is concerned (HAMMEL, 1956). Thus one cannot use thermoregulation to explain their white colour (SVIHLA, 1956). More recent work (e.g. HAMILTON and HEPPNER, 1967; ØRITSLAND, 1970) has emphasized that the problem is far more complicated than previously believed.

On the other hand, shape and position may enable an organism to absorb extra energy (KROG, 1955). Arctic poppies (*Papaver radicatum*) and avens (*Dryas integrifolia*) flowers are parabolic reflectors that concentrate the incoming radiation energy and raise the temperature inside the bloom significantly above the ambient air (KEVAN, 1970, 1972; HOCKING and SHARPLIN, 1965).

Surface covering may enable an organism to obtain a body temperature above ambient. For example, the hairy covering of 'woolly bear' caterpillars (*Byrdia groenlandica*) enables them to retain enough radiant energy to achieve a body temperature of 12°C when the surrounding air is only 3°C (Kevan, in CORBET, 1972). Several times in the High Arctic I have observed 'woolly bears' locomoting over the snow surface when the air temperature was below freezing, but with intense solar radiation. Once, on Devon Island, with the air temperature −3°C, I observed a 'woolly bear' active on the snow surface. Later a snow squall dusted it with a light layer of snow about two flakes thick but sufficient to block incoming energy so that it chilled, lost coordination and could only wriggle helplessly. Thus an analysis of hours of sunshine for any locality may be useful in understanding the bioclimate of some organisms.

Another example may result in survival knowledge for man. Snow cover has a very high reflectance (albedo), even for infrared. On the other hand, the clear night sky is an infinite heat sink; that is, it can absorb heat indefinitely and never reach equilibrium. So a snowshoe hare (*Lepus americanus*), or a red squirrel (*Tamiasciurus hudsonicus*), or a man, can avoid losing heat by infrared radiation to the heat sink of space by putting snow between himself and the night sky. A hare does this by huddling under a shrub that is bent over and covered by *qali* (the snow on trees), a squirrel does this by tunnelling underneath the *api* (the snow on the ground). A man can achieve the same protection by making a snow shelter or 'quinzhee' (ELSNER and PRUITT, 1959; PRUITT, 1973).

## 2.2   Light

Visible light also pulsates annually in boreal regions. At the Arctic Circle, right after the winter solstice, the sun peeks over the southern horizon and then slips back out of sight. Each day it appears a little farther towards the east, rides a little higher in the sky at noon and disappears a little farther towards the west, until by the vernal equinox it rises in the east, rolls through the southern sky and then sets in the west. It continues its northward march day by day until at the summer solstice it swings in a tipped circle around the sky (maximum elevation of 45° above the horizon) and just brushes the northern horizon at midnight. The pattern is retraced until the winter solstice. As one goes north of the Circle, the intervals when the sun is below the southern horizon in winter and above the northern horizon in summer increase dramatically.

Thus the sun never rises very high in the sky in boreal regions and for long periods is just under the horizon. Its rays, then, are rarely excessively strong, although reflection may be severe. Boreal dawns and sunsets are prolonged and spectacular and for long hours the sun shines *under* the canopy of the spruces. The annual pulsation in light and the constantly-changing photoperiod are perhaps the most important differences distinguishing arctic from alpine environments.

The tree line separates two different regimes of energy reflection. In the taiga incoming energy is reflected back and forth and trapped between individual needles until absorbed. In the tundra, because of the lack of structural complexity in the vegetation, more radiant energy can be reflected back to the sky and thus is lost (BILLINGS and MORRIS, 1951).

During the time of rapidly *increasing* photoperiod (and, also, most intense light) plants are inactive. The period of most plant growth is the time of long days, and later, *decreasing* photoperiod. Many boreal plants, therefore are 'long day' plants. That is, their growth, flowering and seed-set are adapted to long photoperiods (SCHWABE, 1956).

Light affects animals in several ways (ROWAN, 1938; PEIPONEN, 1970). Changing intensity and photoperiod govern migratory movements of some species, ripening of gonads and onset of reproduction in others and pelage change in still others. Even in the continuous daylight of the Arctic summer, light intensity decreases at 'night' and bird activity declines accordingly (KARPLUS, 1952). At the highest latitudes birds exhibit flocking behaviour and begin to migrate before the photoperiod actually shortens but while the intensity is decreasing.

It is well known that changing photoperiod governs pelage or plumage change in mammals and birds (NOVIKOV and BLAGODATSKAYA, 1948; SALOMONSEN, 1939–40). Thus the snowshoe hare turns brown to white in response to a particular sequence of light and dark periods, quite independent of the arrival of the complete winter snow cover (LYMAN, 1943). The hare goes through a stage of mottled, irregular colouration, a pattern that matches quite closely the autumn background of patches of new snow, brown needle litter and dark shadows under alder trunks.

Photoperiod is not all-powerful in governing some phases of animal activity. For example, some small rodents, living under a thick snow cover, come into breeding condition long before increasing photoperiod can be detected in their bioclimate. Indeed, some boreal small mammals breed all winter long under the snow cover, governed by the temperature of the subnivean space and amount of protein in their food (KOSHKINA and KHALANSKI, 1962; MYSTERUD, 1966; KHLEBNIKOV, 1970). Even though the photoperiod has lengthened the required amount, other species of small rodents may not breed unless they can obtain a supply of high-protein food (PINTER and NEGUS, 1965). In extreme boreal regions in summer, when the photoperiod does not change for extended periods, some animals abandon circadian rhythms and cue various processes to other environmental factors such as temperature.

## 2.3   The effect of atmospheric moisture on radiant energy

Atmospheric moisture affects radiant energy exchange by refracting, reflecting and absorbing it. In a strongly continental region, well removed from maritime influences, the air is dry and radiant energy may flow inwards (and outwards to space) relatively unimpeded. In winter, boreal regions are usually continental in climate since the snow-covered surface of the polar ice pack reflects radiant energy just as the snow-covered land does. In summer, however, the many ice-free lakes as well as the water-soaked melting snow on the polar pack and the open water of the leads all combine to lessen the continentality of boreal climates. Thus in winter boreal regions have little cloud cover but in summer low clouds and fog are widespread. This condition is particularly noticeable in the Queen Elizabeth Islands and the Polar Basin (HARE, 1970).

## 2.4   Seasons

It is obvious that the temperate-zone concept of seasons cannot apply in regions where the basic physical factors fluctuate so violently. Consequently biologsts have devised various schemes to classify local phenology into time blocks. Some of the factors used to distinguish the categories are type and frequency of snowfall, stages of snow melt, snow-free periods, melting of lake ice, and so on.

# 3 Water, Soil and Permafrost

## 3.1 Water

In general, precipitation decreases northward. Large portions of the boreal regions, especially the continental tundra and the High Arctic, are 'deserts' in terms of actual precipitation. Large areas of the taiga have very little rainfall in spring and early summer.

Fog occurs whenever and wherever there is open water, and is more common in mid- and late summer. Very little is known about the ecological effects of fog in boreal regions, but by extrapolation of the results of research done in temperate coastal mountains, it can be postulated that fog contributes a considerable increment to the annual water budget of a tundra region.

Snow is one factor that unites the boreal regions (RIKHTER, 1962). Because of its importance, discussions of it will occur in several contexts here. Snow begins as tiny ice particles around condensation nuclei in the upper air (MASON, 1961). The air must be supersaturated with water vapour. As the ice particles accumulate more molecules of water they become heavier and start to fall. As they fall they usually pass through air masses with characteristics different from the mass that engendered them. Thus they change, grow larger and more complex, may be partially melted, or buffeted by wind so that what finally falls onto the earth's surface is quite different from the original particle of ice that started to fall (ZAMORSKII, 1955). Once the snow crystal arrives at the surface, whether this be the earth itself, vegetation or other snow crystals, metamorphosis begins. There are three factors which act to change snow crystals—wind, heat and moisture (PRUITT, 1960).

Wind can move the crystals, breaking them apart and eroding them into smaller spicules or dust-like fragments. Unbroken six-armed stellate crystals lie together loosely, supported by the tips of the arms. The needle-like spicules of wind-moved snow fit together snugly in contact for more of their length. Thus a snow cover consisting of wind-moved fragments contains more ice than does a cover of unbroken crystals. The densities of undisturbed taiga snow are in the order of 0.05–0.3 whilst those of wind-consolidated snow are 0.3–0.5 or even more (WILLIAMS and GOLD, 1958).

With time, molecules of water move between crystals and bond them together. Undisturbed crystals bond together only where the arms touch, resulting in a rather fragile structure. Wind-moved needles, on the other hand, bond together over more of their lengths to form a more solid mass.

A snow cover, then, may be considered as an emulsion of ice and air where the amount of air greatly exceeds that of ice. When such an emulsion covers the ground it retards the flow of heat (geothermal or stored) from the earth to the supranivean air. Thus the snow crystals on the very bottom of the cover are warmer than those above them and these, in turn, are warmer than those above them. The air between the snow crystals also exhibits such a thermal gradient. Since warm air can hold more water vapour than cold air the air between the snow crystals exhibits a moisture gradient from the bottom of the cover to the top. Molecules of water flow from the attenuated rays of the warmer crystals and attach themselves to colder crystals.

The bottom layer of snow is not only the warmest—it has been subjected to such molecular loss the longest. Small crystals disappear and large crystals grow at the expense of smaller ones. The resulting material consists of interconnected columns of large (3–10 mm) hollow pyramids that look as if they are made of minute glass logs. This fragile, lattice-like structure is properly termed *pukak* (Fig. 3–1) and is of very great importance in the lives of small mammals and certain invertebrate animals.

In boreal regions there are, then, two main types of snow cover—taiga

**Fig. 3–1**   Vertical profile through tundra upsik. Note lemming tunnels through the pukak layer and reindeer droppings on surface of upsik. North-west of Övre Soppero, Sweden.

or *api* (Fig. 5–2) and tundra or *upsik* (Fig. 3–1). Api varies little in thickness from spot to spot (with certain marked exceptions) within any one region, but it varies greatly in thickness from year to year. Upsik, however, varies greatly in thickness from site to site but is remarkably constant from year to year particularly in the Low Arctic. Duration of snow cover varies greatly, as one would expect, with variation decreasing from south to north. Perhaps the prime characteristic of a boreal region is a non-intermittent snow cover, i.e. one that comes in the fall and remains until spring.

The ecological effects of a snow cover are so multi-faceted and all-pervading that one might accurately state that boreal ecology is the study of the ecology of snow. Briefly, a snow cover affects living things in four ways: it protects ground plants from low temperatures and from desiccation; it accumulates on trees (qali) and may break them or influence their shape; it protects some animals from low temperatures and from predation; it hinders other animals in their movements and food-procuring activities.

The moisture/temperature gradient through a snow cover is the same phenomenon that occurs with any insulation. An understanding of the phenomenon is of considerable importance to human activities in boreal regions.

With cold-weather clothing it is relatively easy to acquire sufficient insulation to control the rate of heat loss for human comfort. Human skin is comfortable through a wide range of temperatures, say, 15° to 32°C, but air at these temperatures will hold much more water vapour than air outside the clothing at, say, −40°C. Moreover, the body produces moisture in widely varying amounts, depending on exercise. Consequently, somewhere in the insulating layer of clothing will be the freezing line where water vapour will precipitate as ice particles. Depending on movement and exercise the frost line will move out from, or in towards, the body. Thus the insulating material must be designed so that the frost particles can be removed by shaking or beating the garment.

The native peoples of boreal regions evolved the 'atiggi' or hooded parka that falls free nearly to the knees so that air can circulate up, be warmed and then escape either through the neck opening or out the bottom again (forced downwards because of the garment's flapping) carrying some moisture with it. Caribou (*Rangifer*) fur is an ideal material for such clothing since the frost particles will fall free when the garment is removed, turned fur-side out and shaken (STEFANSSON, 1945). There are several artificial fur-like materials of nylon or orlon that are suitable for cold-weather clothing. Down garments are quite unsuitable for extended or continuous use in very low temperatures because water particles condense and freeze in the down between the fabric layers and are trapped. A down sleeping bag gains about 100 g of ice crystals per night of use at −40°C resulting in progressive loss of insulation.

The problem of moisture loss is exacerbated in the case of footgear by the need for strength and wear-resistance. Leather, rubber or other non-permeable materials are not suitable for continuous use below −20°C because they trap perspiration inside the shoe where it forms ice crystals.

Undoubtedly the best footgear for bush wear in northern winters are one or two pairs of caribou fur socks inside moosehide moccasins. Such foot coverings are unsuitable for 'modern' conditions when working with oil, petrol, water and grime which soak into the moosehide. Perhaps the best substitutes today are felt boots, with outer rubber soles, reinforced with leather at stress and wear points, and large enough so that one or more pairs of heavy all-wool socks (knitted or felted, no synthetics) and thick felt insoles can be worn.

Since moisture is such a problem in extreme cold, overheating and sweating should be avoided. Flexible thermoregulation can be achieved by the wearing of many thin layers of clothing rather than a few thick layers. Thus, instead of heavy wool underwear, heavy shirt and down parka, a person is more comfortable in fishnet underwear, thin jersey, one or two or even three wool sweaters covered by a thin, tightly-woven windproof shell. Layers can then be peeled off *before* increased activity and sweating, and later replaced.

Such replaceable layers of insulation are not possible in houses or other structures—here the insulation must be isolated completely from the moisture vapour that continually travels along the temperature gradient from warm inside to cold outside. The isolation is achieved by a vapour barrier of foil or plastic on the inside of the insulation. In regions of constantly low winter temperatures a gravity-controlled ventilation system greatly aids the elimination of excess moisture and minimizes thermal stratification inside the structure.

## 3.2   Soil

Soil is an emulsion of finely-divided parent material, organic material, air and water. The size and relative proportions of the particles may vary, resulting in a wide spectrum of soil types. The history of development of the soil influences its properties.

Almost all boreal soils are relatively young since most northern regions are still recovering from glaciation. Consequently, northern soils may not agree with topographic correlations found in southern or non-glaciated regions (DREW and TEDROW, 1962).

Taiga soils, in general, are *podzols*, that is, soils formed under the influence of an acidic conifer needle litter. Such a soil has on its surface a mat of partly-decayed needles and mosses, underlain by a zone of whitish or grey fine-grained material which, in turn, is underlain by a zone of reddish or brown coarser material. The whitish layer is the zone of leaching. The leached materials are deposited in the reddish layer. One

result of the leaching is a strongly acid soil with reduced levels of plant
nutrients in the upper layers.

Tundra soils, because they are churned by frost action, remain
immature (WASHBURN, 1956), i.e. the particles do not remain in the same
relation to each other long enough for the characteristic layers or
'horizons' to develop. Low temperatures hinder decay of organic matter
and the scanty soil fauna cannot redistribute the material in the soil
(TEDROW and CANTLON, 1958; TEDROW, 1970).

A soil transect across the forest-tundra would reveal, not a sudden
change from podzols to non-podzols, but a gradual change in a number
of soil characteristics, particularly the amount of organic matter. The
picture is complicated further by post-Wisconsin (post-Würm) climatic
fluctuations which resulted in the forest-tundra zone shifting back and
forth. In certain regions the present tundra vegetation invaded after the
preceding forest was destroyed by fire or other human influences.

Northern rivers and streams frequently flood in springtime. The
reasons are three-fold: melting occurs rapidly during spring 'break up';
in permafrost regions the soil and organic matter is already water-
soaked; northern drainage patterns are usually immature.

Low temperatures and soils that are water-logged because of faulty
drainage (caused by permafrost) and acidic (because of conifers and
*Sphagnum*) result in few species of soil invertebrates and microorganisms.
Thus the vegetation, when it dies, remains intact and undecayed for very
long periods of time. The resulting mass of water-logged, acidic,
compressed but only slightly-decayed vegetation is *peat* (DEEVEY, 1958). In
boreal regions there are two main peat-forming groups of
plants—*Sphagnum* and sedges. Both types, of course, may have a
considerable percentage of woody material incorporated in the mass.
Peat occurs in four main types of situations: bogs, fens, marshes and
swamps (GORHAM, 1957; POLLETT, 1968).

*Palsa bogs* occur in arctic conditions and consist of a series of steep-sided
mounds, up to six or seven metres high, with frozen cores that may be
extensions of permanently-frozen peat and underlying mineral soil (Fig.
3–2).

Ever since the last continental glacier retreated, peat has accumulated
in bogs, lake bottoms and depressions and, in certain regions, spilt over
to coat even the hills. The energy/material cycles are broken at the
'decomposer' link. The vast areas of peat in boreal regions are sinks for
energy and vegetation, sinks that in many cases have operated since the
first plants invaded the raw landscape left by the melting ice.

It is rather ludicrous to attempt to use such sinks for monetary gain by
investing still more energy and materials—to build machines to refine oil
to run the machines, to ditch and drain the peat so other machines can
plant trees that transfer a minute amount of the peat energy to another,
short-term sink (i.e. wood) and that require expenditure of more energy

to cut, transport and manufacture, not badly-needed food, but paper! Instead of such an engineering approach to a biological problem efforts would be better spent trying to breed or evolve (perhaps with ionizing radiation?) organisms that could continue the interrupted decomposer cycle and release the locked-up nutrients into the biosphere. Perhaps even several steps in a foodchain could be artificially evolved until extant

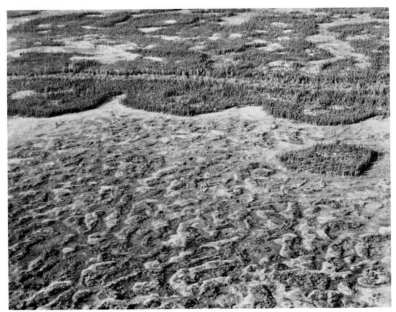

**Fig. 3–2**   Patterned Ground. Palsas, foreground, in peat adjoining a beach ridge of an earlier, higher level of Hudson Bay. South of Cape Tatnam, Manitoba.

organisms could take over and continue the process to a stage man could use. Such a biological solution—investment of little actual energy but harvesting large amounts of energy and/or materials in the form of food (particularly protein)—would be more in line with the world's needs than paper production.

The possibility that boreal peat may be more important in its present form than in any other should not be disregarded. The great quantities of peat in boreal regions act as buffering agents, damping climate oscillations, and as vast sponges, holding water and releasing it in a constant flow.

## 3.3   Permafrost

One of the most important factors in the ecology of northern lands is *permafrost*, by definition—any substratum that is below freezing for two or more years (PRUITT, 1970a). Thus it may be loam, peat, sand, gravel or even solid rock. Permafrost may be continuous, discontinuous or sporadic. The areas of unfrozen ground that are found between areas of permafrost are called *talik*. The *active layer* is the ground above the permafrost which is subject to annual freeze and thaw. Most tundra has continuous permafrost, but there are exceptions such as maritime tundra in eastern Canada.

Permafrost, can never be considered alone, either in its genesis and degradation or in its ecological effects. Permafrost forms a complex, dynamically-balanced system with climatic history, vegetational history, recent climate, present vegetation, substratum, topography, solifluction and animal activity all interacting. If any one of the factors changes, the permafrost regime shifts.

In spite of the fact that permafrost affects approximately one-fifth of the land area of the northern hemisphere surprisingly little is known about it. Permafrost has a profound effect on vegetation, particularly tree-like vegetation. Consequently most distribution maps of permafrost, particularly those delimiting the discontinuous and the continuous, are based to a large extent on plant distributions (Fig. 1–1). Thus circular reasoning must be guarded against in the interpretation of factors influencing plant distribution.

Permafrost forms whenever there is even a small negative heat imbalance each year, so that more heat flows away from the substratum than flows into it. Such a situation will result in a thin layer being added annually to the permafrost (BROWN, 1966b). Numerous climatic factors affect the genesis of permafrost—snow cover, cloudiness, continentality, duration and severity of seasonal extremes, etc. The most important climatic factor in the distribution of permafrost is the mean annual air temperature. The southern limit of permafrost coincides roughly with the −1°C mean annual air isotherm (BROWN, 1960 and 1966a). The climatic factor of next importance is snow cover. ANNERSTEN (1964) has shown that the critical thickness of snow cover for permafrost to survive is 40 cm.

Vegetation cover is also an important factor in the permafrost regime. Vegetation acts as an insulator; consequently the thicker the vegetation mat the more insulation. In the High Arctic where the vegetative cover is incomplete it has minimal effect on the permafrost regime. In the Low Arctic, where the vegetative mat is complete, it governs the thickness of the annual thaw layer. If the vegetation (i.e. the insulation) is disturbed then thawing increases. A thaw-erosion-thaw cycle then begins and thermokarst may result. (Thermokarst refers to the cavities caused by melting and subsequent shrinkage in volume of soil or subsoil that had a high ice content.) In the forest-tundra there is more vegetation and more snow accumulation. Consequently the permafrost temperatures are not too far below freezing. Disturbance to vegetation is critical and thermokarst is a constant problem. In the taiga vegetation achieves its greatest influence in the permafrost regime. Even slight disturbances may result in severe degradation of permafrost (PÉWÉ, 1957; BROWN, 1966b and 1970).

The insulating effect of vegetation and its seasonal variations are very important in permafrost formation on the southern edge of the permafrost region. In general, moss and peat are more important than the shrub and tree layer, although VIERECK (1965) has demonstrated a complex cycle resulting from the effects of trees and *qamaniq* (bowl-shaped depression in api under spruce branches) (PRUITT, 1957 and 1959b).

Relief may affect permafrost through orientation and degree of slope. This is particularly evident in mountainous regions but it is also effective on a much smaller scale, e.g. there is a different permafrost regime on north and south-facing stream banks and on the north and south sides of peat mounds (BROWN, 1970; SALMI, 1970). In turn, permafrost affects relief through thermokarst and solifluction (the movement downhill of water-saturated soil or parent material by freeze-thaw processes).

Any change that results in a positive heat balance will be favourable to permafrost degradation. Thus, a climatic change that causes more winter cloudiness and more summer sunshine may result in a more positive heat balance in the substratum, with accompanying degradation of the permafrost. Likewise, any change in vegetation will undoubtedly cause changes in the permafrost regime.

In general, man's activities in regions with boreal vegetation lessen its effectiveness as insulation of the substratum and act almost invariably to degrade permafrost. Over wide stretches of the boreal regions the permafrost is in such delicate climatic balance that even apparently minor disturbances may have profound, even disastrous, consequences. In the Ogotoruk Valley, north-western Alaska, which was subjected to intensive traffic in the period 1959–61 by 'weasels' and other relatively light, track-laying vehicles especially designed for Arctic use, some of the trails have now thawed and eroded three or four metres in depth. On the Arctic slope of Alaska trails made by winter tractor trains in the late nineteen-forties have thawed, eroded and caused gullies ten or so metres wide, three to five metres deep and many kilometres long (HOK, 1969).

There is almost always an unfrozen zone beneath water bodies that themselves do not freeze solid annually. Thus, a formation of a new water body or increase in size of an already-existing water body, because of damming or flooding will result in degradation of existing permafrost. The resulting settling and thermokarst may be of profound ecological (and economic) importance.

It is, at present, impossible to distinguish with certainty between permafrost-caused effects on the substratum and those caused by annual freeze-thaw. Many workers have noted such physical effects as immobility of the substratum, patterned ground (Fig. 3–3) (WASHBURN, 1956; DREW and TEDROW, 1962), heaving (Fig. 3–4) (FRASER, 1959), palsas (Fig. 3.2) (SALMI,

1970), pingos (pressure-raised conical hills in old lake sediments), oriented lakes and thaw lakes. In general the patterns caused by frost-heaving are symmetrical on level ground and become progressively more elongated the steeper the slope. Two of the basic causes of patterning are: (1) differential freezing and thawing and (2) differential movement of particles.

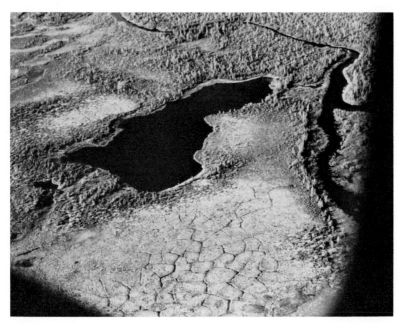

**Fig. 3–3**   Patterned Ground. Soil Polygons. North-west of York Factory, Manitoba.

BRITTON (1957) noted that ice wedge polygons require permafrost. Contraction cracks open in permafrost at low temperatures and become filled with hoar and refrozen melt water. Repeated annually, increments may be 0.5–1.5 mm/year, and eventually may result in wedges up to 8 or 9 m wide. The surface manifestation of the pattern depends mainly upon differential rates of thaw between ice and the matrix of frozen mineral and organic material. If the surface insulation is equally distributed the thaw will cause the ice level to be lowered and a polygonal ditch system results (Fig. 3–3). In this ditch *Sphagnum* grows and accumulates and may transform the ditch to a dyke.

DRURY (1956) and GILL (1975) have postulated some exceedingly interesting natural relationships between vegetation and permafrost. One of the prime references to vegetation-permafrost interactions is TYRTIKOV (1959). Undoubtedly the most direct effect of permafrost is upon plants by chilling the soil and retarding general growth. Of course, the vegetation acts, in turn, to chill the soil by shielding it from incoming solar radiant energy. Permafrost prevents deep root penetration, so that plants growing above a shallow layer are relatively unstable. On the other hand, permafrost acts as a reservoir of considerable heat and thus acts to damp wide fluctuations in soil temperature.

Because of its impermeability, permafrost has a great influence on runoff. Permafrost causes a more stable moisture regime by gradually releasing water. These two factors combine to cause permafrost areas to be characteristically wet and boggy. In spite of an excess of water, the soil is physiologically dry for plants because the low temperature of the water (and its frequent low pH due to an excess of organic solutes) retards its absorption by roots.

Because the active layer is a relatively thin film over the permanently frozen mass, minerals in the active layer may be exhausted and no replacements are possible from the store immobilized below. Mineral cycling is thus restricted to the active layer. Understanding of this problem is compounded by the difficulty of distinguishing between the effects of permafrost and those of the annual freeze.

The upper boundary of the permanently frozen mass is a sensitive indicator of ecological variations and moves upward or downward. Such pulsations result in widely varying site characteristics for plants. When the plants, particularly trees, have varying health they are subject to wind tipping, qali breakage and varying growth rates. Such variations combine to produce twisted and stair-step main stems or 'drunken forest' (Fig. 3–4). Soil heaving related to annual freezing is also an important factor here. On irregular terrain, solifluction and downslope mass movements may have significant effects on the biota.

One important biological problem about which relatively little is known is the restriction of aeration due to permafrost. Soil microorganisms and their activity thus may be different from those in non-permafrost regions and the intensity of their activity reduced.

So far the concern has been with the interrelations of permafrost and plants. Permafrost also has profound, if less well-known, effects on animals. For example, the vegetational factors that dampen temperature oscillations throughout the year create conditions that are also thermally agreeable to small mammals and soil invertebrates. The reservoir of heat in a mass of permafrost is useful to animals as well as plants. But just as plant roots cannot penetrate permafrost, neither can animals. Thus fossorial animals, particularly mammals, are notably absent from permafrost areas. FLOROV (1952) has shown that permafrost may also influence the distribution of an insect that has a burrowing larval stage.

On the other hand, the reservoir of moisture in permafrost that is so important to plants may act against many animals. Small mammals, in particular, are very susceptible to wetting. Saturation of the active layer,

**Fig. 3–4** Thinly-stocked subarctic taiga, 'drunken forest' of black spruce, caused by frost-heaving. Old John Lake, north-eastern Alaska.

and especially the presence of *Sphagnum*, frequently results in few or no small mammals. Saturation of the active layer requires special adaptations by large, heavy animals. The large hoofs and free-swinging gait of the caribou are undoubtedly adaptations for snow cover but also serve as flotation mechanisms on boggy or water-soaked substrata. Moose (*Alces*) are particularly well-adapted to traverse tangled, lattice-like vegetation. Such a vegetational configuration may occur as a result of trees and shrubs being twisted or killed by permafrost or solifluction. The Arctic ground squirrel (*Citellus*) is an anomaly in that it is a hibernator successfully existing in an environment notably unsuited for hibernators. These animals construct their burrow systems, especially their hibernacula, in situations where the permafrost line lies below them. The fact that soil nutrients may be immobilized in permafrost must have a profound effect on tundra animals as well as the plants on which they feed.

Palsas are a frequent accompaniment of permafrost. They provide dry islands in a saturated sea of peat (Fig. 3–2) and thus enable some small mammals to survive there. For example, the 'heather vole' (*Phenacomys*) appears, in parts of its range, to be limited to palsas and palsa-like formations (FOSTER, 1961). Palsas may also become important winter feeding places for caribou for two reasons—the dry surface supports better lichen growth than the boggy peat surrounding it and also the exposed, elevated surface is more likely to have a thinner snow cover than the peaty lowland.

The climate that causes permafrost to form also causes an annual ice cover of lakes, rivers and protected marine channels. BANFIELD (1954) has discussed the important role that ice plays in the distribution of some northern mammals.

Animals and plants can scarcely be considered as separate entities in ecology. This is particularly true in boreal regions. Biotic communities over permafrost sometimes have, to temperate-zone eyes, strange relationships. It has been pointed out by BRITTON (1957) that because of erosion during polygon cycles different communities may be in contact. This is not necessarily an indication of their successional relationship.

VIERECK (1970) described how changes in plant cover can result in changes in soil temperature. He stressed the insulating effect of an accumulated moss or organic layer. Since saturated frozen peat has three times the thermal conductivity of dry peat, during warm periods in summer when the upper layers dry out they decrease the heat flow downward. In fall the organic layer becomes saturated. When frozen in winter it becomes an even better thermal conductor, but for heat flowing upward. Thus a negative heat balance develops in the soil (BROWN, 1966b). Successional development of vegetation, especially the development of a thick moss layer, results in colder soils and eventually permafrost. Frozen soil prevents water percolation, resulting in wetter soils and inhibition of tree growth.

DRURY (1956) described a way in which bogs may form by thawing of permafrost. First there is a break in the moss on the forest floor. The break could be caused by fire, blown-down trees, game trail, qali (PRUITT, 1958) or bursting of a pingo. However the break occurs, thawing starts. This results in slumping and a water-filled depression which grows in size. *Sphagnum* and sedges invade the depression. As the bog pool grows it is invaded by other mat-forming vegetation and thus the bog grows at the expense of the forest. Drury postulated a cycle, with bogs advancing by thawing, followed by organic deposition in the wet depressions, followed by progressive drying as the peat and silt deposits thicken. Eventually a black spruce-tamarack forest is re-established, accompanied by a rise in the permafrost table.

On the Arctic Slope of Alaska the degraded permafrost below the ruts and gullies caused by tractor trains supports hydrophilous vegetation (HOK, 1969). Populations of the 'Alaska vole' (*Microtus oeconomus*) have spread from their normal riverine locations into the bottoms of these ruts and thence into the drier uplands. Thus this species was brought into extended contact with collared lemmings (*Dicrostonyx groenlandicus*). Moreover, one can only speculate on the possible genetic results of having a population attenuated to one or two *Microtus* wide and perhaps thousands of *Microtus* long.

It can be seen, then, that permafrost is one of the major components of the environment of tundra biotic communities. Its effects are towards both stability and change in the community. Human activity tends to degrade permafrost. In permafrost regions any human activity, especially one to be accompanied by modification of natural vegetation associations, should, therefore, be preceded by investigations to determine the changes to be anticipated in the permafrost regime. The ecological repercussions of such changes must be evaluated, of course, in relation to the advantages of the proposed human activity.

# 4 Bioclimate

Energy from the sun is absorbed by the substratum, heating it. When the substratum becomes hot enough the air immediately above it becomes turbulent. At some level above the surface the turbulence suddenly declines and ceases, being 'swamped' by wind. The region above the zone of turbulence is technically the *macroenvironment* while the zone of turbulence itself is the *microenvironment* (GEIGER, 1965). The zone of turbulence occurs when the sun is between 10° and 30° above the horizon (CORBET, 1972).

Another term of considerable conceptual importance is *bioclimate* which refers to the total complex of physical factors impinging on an organism where it is at that particular time, or the total during an organism's life and movements.

A red squirrel, for example, in the taiga in summer lives in air that varies from, perhaps 32° to 10°C. Later in the year the range of bioclimate temperatures might be 0° to −20°. In mid and late winter the squirrel will be exposed to perhaps −20° to −30°. But whenever the ambient temperature falls below about −32° the squirrels remain in their subnivean nests or are active only in the pukak space, at temperatures of 0° to about −10° (PRUITT and LUCIER, 1958). Thus, the range of bioclimate temperatures for the red squirrel extends from +32° to −32°C. Consequently, red squirrels never experience a significant portion of the taiga environment and lack adaptations to it.

On the other hand, the bioclimate of a spruce tree extends from the conditions experienced by the deepest root to those experienced by the growing tip high in the air. In terms of temperature, for example, the root lives in an environment that changes little from day to day and not much from season to season. Indeed, because of the lag in heat penetration of the soil, the temperature around the deepest root may be quite out of phase with that of the growing tip. Moreover, the trunk and growing tip protrude through the microenvironment into the macroenvironment.

The temperature bioclimate has been mentioned only because it is relatively easy to measure and because so many data are available, but it should be remembered that heat content and heat transfer are the basic thermal parameters. Other factors of the bioclimate exist, such as light, moisture, air movement, evaporation, and so on. It is clear that in order to describe the bioclimate of any species a thorough knowledge of its natural history is needed; its physical position, orientation, movements and cycle of activity and torpor.

## 4.1   Taiga

In order to understand the microenvironment of the taiga an under-standing of its macroenvironment is required (Figs. 4–1 and 5–2). The ambient temperature fluctuates annually, from a high of, say, +35°C to a low of −50°, −55° or even lower. The greater spread between short-term maxima and minima in summer than in winter should be noted, and the great seasonal pulsation in light, the exact sequence varying with latitude, recalled. Sometimes the air is filled with precipitation, rain, fog or snow, which may be swirled by wind. The atmosphere varies from saturation with moisture to very dry, although most taiga regions exhibit dry air for long periods of the year. Wind, and the noises it produces, occur at irregular intervals.

In contrast, the floor of the taiga can be considered as microenviron-ment, from among the tops of the moss plants ('surface') to 23 cm below the surface ('−23'). This physical region is the habitat of voles and shrews and a host of invertebrates, as well as most plants at some stage in their life cycle. Thus a survey of this microenvironment is also a survey of their bioclimate.

In summer, the ambient air is generally warmer than the forest floor, considerably warmer than at −23 cm. In the autumn the ambient temperature falls until the air is cooler than the surface. The time of uniformity is the 'fall thermal overturn'. Air temperatures continue to fall, in close agreement with the floor temperatures. Over a period of time in the autumn the snow cover arrives, the actual times varying with the sequence of meteorological events that bring moist air into contact with cold air. The floor temperatures continue to follow closely the flunctuations in air temperatures until the snow cover reaches 15 or 20 cm thickness, then they tend to stabilize and may even rise. The 15 or 20 cm thickness can be called the 'hiemal threshold' because it is the true beginning of winter as far as the small creatures of the forest floor are concerned. The time between the fall overturn and the hiemal threshold is the 'fall critical period' since the greatest thermal and survival stresses occur then. In regions with a denser snow cover that has a greater heat transmission the hiemal threshold must be thicker. Some regions, especially the High Arctic, have 'true winter' in a patchy pattern corresponding to zaboi accumulation (see p. 61).

The heat from the earth has two sources—geothermal heat from the molten core and solar heat which has penetrated during the summer. This heat flows to the surface and during the winter dark period is almost the only source of microenvironmental heat. Thus the base of the snow cover is warmer than the upper layers.

During the period of snow cover (Fig. 4–1) the −15 cm and −23 cm levels are largely free from fluctuations. The surface and −7.6 cm levels fluctuate in approximate agreement with ambient temperatures but with

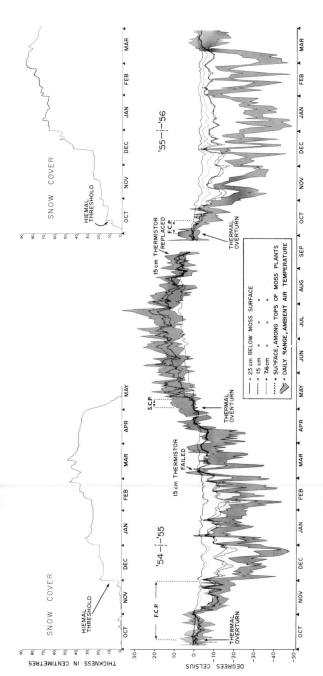

**Fig. 4–1** Some environmental characteristics of subarctic white spruce (*Picea glauca*) taiga, central Alaska (65° 50′ N. Lat.). Points to note: variations in length of Fall Critical Period (FCP) in different years; variation in api thickness (reflected in severity of temperature bioclimate of soil-inhabiting organisms). See text for complete discussion. (Originally published in *Arctic*, **10**, 1957, reproduced with permission of the Arctic Institute of North America.)

markedly less amplitude and far warmer (by as much as 35 Celsius degrees). Contrast this situation with that of summer when the surface and −7.6 cm levels are within the range of daily extremes. Note also that the −23 cm level remains farther above the winter ambient than it remains below the summer ambient. During the period of snow cover, then, the bioclimate of taiga voles and shrews is relatively mild (temperature extremes of 0° to −23°C) and stable.

In this subnivean bioclimate the air is saturated with moisture, making ideal conditions for shrews which have poorly-developed moisture-control mechanisms. A saturated atmosphere also offers ideal conditions for communication by scents. The subnivean bioclimate is dark, because a snow cover 50 cm thick transmits only 1% of the incident light (GEIGER, 1965). Moreover, the light that does penetrate is in the longer wavelengths (red) that are not detected by some voles (*Clethrionomys* and undoubtedly others). Thus the thickness of the snow cover and its light-transmission may be a factor influencing breeding in subnivean mammals (EVERDEN and FULLER, 1972).

The subnivean bioclimate of voles and shrews is silent, with no wind. Sound insulation from the supranivean environment is virtually complete; all that can be heard of a howling storm is the creaking of tree roots in the soil (PRUITT, 1967). Footsteps cannot be heard until they are only a metre or so away but, conversely, nothing can approach without making a noise.

Because the snow surface reflects almost all incoming radiation, and the low rate of heat transmission retards heat flow from lower layers, the topmost few millimetres of the snow cover and the air immediately above it become very cold. Such a frigid layer may be of extreme importance to animals on the snow surface, e.g. snowshoe hares or small birds foraging for fallen birch seeds.

In the spring the snow cover disappears by sublimation and melting. During the melt density increases and thus so does heat transmission. The normal temperature gradient from soil to snow surface is destroyed and for a period of time subnivean plants and animals are exposed to thermal stress. The spring critical period extends from the time when the snow cover ceases to protect the forest floor from ambient fluctuations until the time of the spring thermal overturn, after the snow cover has disappeared.

It is clear that the microenvironment of the taiga is notably different from the macroenvironment, particularly between the fall and spring overturns. It is also clear that the single most important factor affecting the bioclimate of organisms on and under the forest floor is the snow cover. While the snow cover is present it acts as an ecotone separating two drastically different environments (Fig. 4–1).

## 4.2   Tundra

The tree line separates environments that are superficially quite
different; upon analysis it can be seen that they differ in degree but not in
type. The tundra macroenvironment, because of the almost-constant
winds, reaches closer to the ground. The tundra microenvironment is
fragmented, with patches of macroenvironmental conditions impinging
onto the convex and protruding surfaces; microenvironmental
conditions are limited to concavities and protected sides of gullies, banks
or rocks.

The tundra macroenvironment (Figs. 1–3 and 1–4), especially the Low
Arctic, is dominated by wind that stirs the air and prevents cold air
drainage and puddling, that boosts evaporative cooling and windchill
(SIPLE and PASSEL, 1945; COURT, 1948), that drives the fog into every nook
and crack, that blasts the snow crystal fragments and the scouring sand
or dust, and that moans and whistles over frost cracks and tussocks.

Tundra macroenvironmental temperatures rarely reach the extremes
that occur in the taiga. In most tundra regions there is usually no more
than a week or two of really 'hot' summer weather. Such a single peak of
maximum temperature may affect the biology and distribution of a
number of organisms, particularly ectotherms. The annual cycle of
temperature in tundra regions shows fewer deviations from a regular
progression than do taiga regions. Thus tundra has fewer freeze-thaw
cycles than do more southern regions (FRASER, 1959). In general, tundra
regions have moist air in summer and dry air in winter. Even though the
air be moist, evaporation and evaporative cooling may be high because of
wind.

Snow cover arrives and builds up in a more regular and predictable
sequence in tundra than in taiga. Again, wind is the shaping influence
here, by sweeping the snow from convex surfaces and depositing it in
concavities. Thus concavities may have sufficient snow cover (*zaboi*) to pass
the hiemal threshold early each autumn while a few metres distant convex
ground surfaces (*vyduvi*) never acquire this protection (Fig. 1–3). An
important aspect of tundra microenvironment is the predictable annual
repetition of such patterning of the snow cover. Densification of the
tundra snow cover by wind (formation of *upsik*) increases its
transmissivity to light and heat, e.g. doubling the density of snow
increases the rate of heat transmission four times (GEIGER, 1965). Thus,
tundra subnivean organisms are not only markedly restricted in
overwintering areas but are also subjected to lower and more fluctuating
bioclimate temperatures than are taiga subnivean organisms. Because
light transmission is high, tundra subnivean organisms may be exposed
to increasing photoperiod in spring at the same time that supranivean
organisms are.

Whereas taiga subnivean organisms live in conditions that are quite different from the macroenvironment, those of the tundra live in conditions that still are very reminiscent of the supranivean environment. The tundra subnivean environment differs from the supranivean primarily in having a saturated atmosphere and in the lack of wind. The snow cover of the tundra, in addition to being highly variable in thickness or even discontinuous, is denser and harder than that of the taiga or forest-tundra. The pukak layer (Fig. 3–1) is less developed than in the taiga. Thus tundra subnivean mammals must expend more energy moving from place to place than do those in the taiga.

# 5 Characteristics of Boreal Vegetation

## 5.1 Adaptations

Plants, unlike animals, cannot move to a better site, they cannot avoid the short cool summer, the aridity, the low levels of organic matter and nitrogen in the soil, the possible presence of poisonous salts of calcium and other elements, the abrasions suffered in winter by their supranivean parts and the peculiar photoperiod (SAVILE, 1972). Tundra plants have evolved effective morphological, physiological and phenological mechanisms to cope with such inimical factors.

(1) *Morphology* Tundra plants, in general, avoid the macroenvironment and supranivean environment by habitat 'selection'. They tend to be low and compact, many are dwarf shrubs. Some form cushions or rosettes, thereby effectively lowering their surface area. Such a shape traps radiant energy by re-radiating part of the energy to leaves and other parts close by. Many consist of elongated creeping stems from which spring vertical leafing or flowering twigs. Tundra plants frequently have thick or waxy cuticles; some are pubescent, being covered with a good imitation of fur.

(2) *Physiology* Tundra plants have adapted to the short and cool growing season by evolving methods to take advantage of every bit of time possible. Because of the lag in the seasons most of them use only about half of the annual amount of sunlight. Some of them sometimes start growth under the snow (KIL'DYUSHEVSKII, 1956; YASHINA, 1961) or before the air temperature is above freezing. Indeed, in the High Arctic purple saxifrage (*Saxifraga oppositifolia*) may be in flower before the 'official' daily mean is above 0°C. Tundra plants not only metabolize at lower temperatures than temperate zone plants, but also can withstand freezing and can resume growth again when thawed.

(3) *Phenology* Most tundra plants are perennials, a category of plant peculiarly adapted for tundra conditions since the flowering and vegetative buds are produced in the fall. The plant is ready, then, to begin growth in spring. Perennials store food—carbohydrates in underground parts, lipids in prostrate stems—so they can grow in early spring before photosynthesis produces new food. In early spring their respiration exceeds photosynthesis while later in the summer the situation is reversed. The breeding systems characteristic of tundra plants combine to produce organisms adapted to reproduce quickly and disperse readily (MOSQUIN, 1966).

## 5.2   Tundra

The appreciation of boreal vegetation has suffered from two phenomena—an excess of hasty, superficial surveys of the tundra and a neglect of the taiga except for exploitation-oriented research.

Tundra plant associations are identifiable, in the main, by differential frequency of the members of the small, specialized flora, not so much by presence and absence of species. Of the many associations, all characterized by the lack of trees, of course, some are easy to delimit while others are more subtle (ALEXANDROVA, 1970). There are three major categories of associations.

(1)   Wetlands, dominated by grasses, sedges and rushes. Sometimes large regions are of these associations, e.g. the Arctic coastal plain of Alaska, encompassing some 60 000 km² has a slope of only about 1 cm per km and is from 50 to 90% ponds and lakes. Great areas of this region are almost pure stands of marsh grass (*Dupontia fischeri*).

(2)   Drier, but still mesic, areas may be clothed with cotton sedge (*Eriophorum*) tussocks, interspersed with dwarf birch (*Betula glandulosa*), ericaceous shrubs, lichens and mosses. These areas encompass topographic types from slight relief to markedly rolling and thus offer a wide variety of types of snow cover. Some of the most important areas for tundra mammals are *Eriophorum*-birch-heath tussocks on rolling uplands (PRUITT, 1970b) (Fig. 1–3).

(3)   Fellfields. In the Low Arctic the tops of ridges are dry, windswept deserts, quite different environments from the relatively lush valleys. The wind-eroded, frost-heaved rock fragments, with crustose lichens on the fragments, with mats of black *Alectoria* lichen and colonies of *Dryas* on the fines (smallest rock particles) make up the 'fellfield'. As one goes northward the fellfield type of habitat assumes greater importance until in the High Arctic it is dominant.

Plants, quite successful plants, are found as far north as land goes, but in the High Arctic a large proportion of the substratum is bare (Fig. 1–4). Indeed, there are some areas with no angiosperms at all, just occasional lichens on the frost-shattered rock fragments. The meaning of 'Arctic Desert' is emphasized when, in the midst of such an expanse of frost-patterned rocks and bare soil, a watercourse or seepage area is encountered; here plants thrive and form a complete cover, a veritable oasis.

In regions with more precipitation, with soils containing more nitrogen and organic matter, as well as in more southerly regions, the vegetative cover spreads from wet areas to clothe upland and rolling areas. Farther south the vegetative cover spreads over all the landscape and increases in height and complexity. Here, in the Low Arctic, shrubs grow tall and, in favoured and protected sites, may form 'fairy forests' with a functional canopy.

Aŋmaŋa

**Fig. 5–1**   Taiga of Scots pine (*Pinus sylvestris*) and Norway spruce (*Picea excelsior*). Note the aŋmaŋa around the hay shed and the fox tracks leading towards it. Oulanka National Park, north-eastern Finland.

What is a shrub and what is a tree? This seems a simple question but actually it is not easy to answer. One definition after another, when tested, fails somewhere in some part of the boreal regions. Since the most rigorous part of this environment is immediately above the snow surface, a tree that grows through this zone successfully is likely to be, eventually, much taller. Consequently, a two-metre minimum height above the snow line, as well as possession of a single central trunk is a good working definition of a boreal tree.

A firm definition is important in boreal ecology because the basic dichotomy in boreal vegetation is the distinction between taiga and tundra (HUSTICH, 1953; TIKHOMIROV, 1970).

5.3   **Forest-tundra**

From wherever the first specimen of a tree species grows, southward to the taiga proper is a transition zone or ecotone of varying width, sometimes as much as 100 km or more. This is the forest-tundra (lyeso-tundra) (Figs. 1–5, 1–6 and 1–7). Within the forest-tundra one may distinguish several 'lines' (HUSTICH, 1966). *Timber line* is the economic forest line, the limit beyond which commercial cutting of trees endangers natural reproduction. *Physiognomic forest line* is the limit of the forest itself, usually where it reproduces itself only in 'good' years. The *tree line* is the absolute poleward limit of trees, and is formed by different species in different areas (HUSTICH, 1953). The *species line* is the outermost limit of tree species, regardless of whether they grow as trees or shrubs or krumholz (wind-pruned plant, usually of a tree species, reduced to a dwarfed, dense shrub). The lines are but expressions of various environmental factors and biological reactions to them. Indeed, as ROWE (1966) pointed out, there are at least 75 tree lines in Ontario alone.

**Fig. 5–2**   Subarctic spruce taiga in winter. Environmental conditions on this plot are charted in Fig. 4–1.

The type of bedrock and resultant soil dictates to a large degree which species will occur in a region or what their local pattern of distribution will be. Peat accumulations are also effective in limiting tree growth. White spruce (*Picea glauca*) (Fig. 5–2) occurs mainly on rich clay and alluvial soils while black spruce (*P. mariana*) (Fig. 1–5) occurs mainly on acid soils or those where permafrost is fairly close to the surface. In the taiga of interior Alaska white spruce occurs on south-facing loess slopes lacking permafrost while black spruce occurs in the water-logged peaty valleys and north-facing slopes underlain by permafrost. The accumulation of peat may, in time, modify the influence of the bedrock. For instance, in the Hudson Bay Lowlands HUSTICH (1966) postulated that the present black spruce on old peat soils had superseded earlier vegetation of larch or white spruce.

Permafrost exerts differential control over tree species. HOPKINS and SIGAFOOS (1951) noted that, in Alaska, tall willows and isolated pure stands of balsam poplar on river flood plains generally indicated the presence of unfrozen ground, while the minimum thickness of the active layer under tall willows on river flood plains was about 2.5–3 m, under pure stands of mature aspen or white birch it was 1–1.5 m, and under mixed white spruce and balsam poplar it was 1 m. In contrast, black spruce grew where the active layer was only 30 cm thick. Frost heaving, slumping and other types of earth movement associated with either annual freezing or permafrost are important in governing growth of tree species or in aiding the development of 'drunken forest' (Fig. 3–4).

## 5.4   Taiga

Since most of the taiga lies in the zone of discontinuous permafrost (Fig. 1–1), and also encompasses several geological provinces, one would anticipate a complex vegetation with a variety of different communities. The taiga does have more species than the tundra, but still remarkably few. In some regions, hundreds of square kilometres will be dominated by four species of trees—the two spruces, birch and aspen. In a few regions literally thousands of square kilometres will have perhaps 90% of the tree biomass in one species—larch.

Because the dominant trees are conifers, particularly spruce, the general aspect of the taiga is essentially the same wherever it is found—spire-like spruces against the skyline, with feathery larches, white-barked birches, dense stands of alders; together forming a mosaic pattern when viewed from the air. The climate, soils, plants and animals form an interacting fabric which is distinct from other associations adjacent to it. Although there are a number of important regional variations to the taiga, they are but phases of what is obviously the same biotic association (Figs. 5–1 and 5–2).

Being composed primarily of highly resinous coniferous trees, the taiga is extremely susceptible to fire. Although wildfire was a regular occurrence in pre-whiteman times, since the taiga was invaded by European culture the frequency of wildfire has increased enormously (SCOTTER, 1964). In interior Alaska about three-quarters of the spruce-birch taiga has burned and in other regions of North America such as northern Saskatchewan virtually all the spruce taiga has burned. In Alaska the taiga characteristically regenerates through stages of fireweed

(*Epilobium*), birch and aspen (*Populus tremuloides*) but on the Canadian Shield after spruce taiga is destroyed by fire jackpine (*Pinus divaricata*) stages persist for many, many years.

The classic concept of a climatic climax vegetational association breaks down in the boreal regions (CHURCHILL and HANSON, 1958). The species making up the mature taiga, indeed, the very life-form itself, may change due to the powerful forces of permafrost, annual freezing and thawing and snow.

An important aspect of taiga vegetation is the presence of berry-bearing shrubs. Such plants as the blueberry (*Vaccinium*), high-bush cranberry (*Viburnum*), low-bush cranberry (*Vaccinium Vitis-idea*), crowberry (*Empetrum*), strawberry (*Fragaria*), cloudberry (*Rubus chamaemorus*) and rowan (*Sorbus*) are famous. The production of berries, and its fluctuations, is important in the lives of taiga animals.

The ground cover of well-developed northern taiga has a high percentage of lichens. In some aspects of the taiga the ground cover is primarily lichen but in closed taiga the ground cover is primarily feather-mosses.

# 6 Boreal Animals

Animals that depend on external sources of heat energy for their enzyme systems are called *ectotherms*. Animals that produce an excess of metabolic heat are called *endotherms*. According to this classification by energy relations some animals fit sometimes into one category and sometimes into the other. Thus Kevan (in CORBET, 1972) and KEVAN and SHORTHOUSE (1970) demonstrated that several High Arctic insects can soak up enough radiant energy in direct sunshine to enable them to operate as endotherms for some time. CHURCH (1960) showed that tundra bumblebees (*Bombus* spp.) raise their body temperature by means of muscle activity and maintain the body heat by their fuzzy, fur-like coat.

Such examples are rare; virtually all terrestrial invertebrates are obligate ectotherms. Thus no taxon lacking a cold-resistant stage in its life-cycle can ecize, that is, invade and successfully maintain a population in, boreal regions (SALT, 1961). Consequently the kinds of invertebrates in boreal regions are few and specialized, but, none the less, are poorly known (WARNECKE, 1958).

## 6.1 Insects

The insects are the best-known group of boreal invertebrates. Because some species affect man's comfort and well-being most of the research effort has been directed at a dozen or so species. Much work, including a lot of old-fashioned descriptive natural history, needs to be done.

Most temperate regions possess several thousand insect species. In contrast, there are less than 600 species of all insects in the vast territory of Greenland. Only 230 species have been recorded from northern Ellesmere Island (DOWNES, 1964).

In most temperate regions the beetles are the most common insect type, making up about half of the total insect species with Diptera about 20% of the total. In Arctic tundra less than 10% of the species are beetles while the Diptera increase dramatically to about 50% for the tundra in general and to 70% for Spitzbergen. At Isachsen (Ellef Ringnes Island) out of 28 species of true insects, 25 are Diptera and, of these, 17 are Chironomidae (DOWNES, 1965). These Arctic insects are most closely related to those of the Temperate regions. There is little endemism.

The insects which have attracted most attention because of their effects on man are in the Diptera and include members of the families Culicidae (mosquitoes), Heleidae (no-see-ums), Simuliidae (blackflies), Leptidae (snipeflies) and Tabanidae (deerflies).

Mosquitoes seek a blood meal in order to obtain the extra energy needed to develop eggs. Some species regularly obtain their extra energy from flower nectar, others can, apparently, lay a reduced number of eggs by digesting their own flight muscles or a special fat body. The literature is replete with conflicting statements about blood meals and nectar meals; clearly much careful research is needed.

Northern mosquitoes are specialized in that they show only two types of life cycles (FROHNE, 1956). The common type of northern life cycle has the animal surviving the unfavourable winter period in the egg stage. The larvae hatch the following spring or summer and the adults emerge.

The other type of northern life cycle has the animal surviving the unfavourable winter period as a hibernating adult. The eggs are laid, hatch, pupate and the adult emerges in one season, without diapause. The adult female mates the first summer but she does not seek a blood meal until after the obligatory hibernation period, then she does so and lays her eggs.

Horrendous are the tales passed on about the viciousness and numbers of northern mosquitoes. SETON (1911), with his insistence on numerical data, was the first to provide definite comparative information on the subject. He counted the number of mosquitoes biting his bare hand in a five-second period. As he travelled north his index rose from 5–10 to 15–20, then on Great Slave Lake in late July, 50–60 and, on the 'Arctic Prairies' above Artillery Lake, NWT, 100–125. HOPLA (1964–5) devised an index to compare relative attraction of different species of animals for mosquitoes. His index was the number of mosquitoes actually feeding on 337.5 cm$^2$ of human forearm in 15 minutes. In the taiga of interior Alaska Hopla recorded a 'biting index' of 54–69 on man and 24–32 on rabbit. On the Low Arctic tundra of northern Alaska he recorded his maximum index, 253. Seton's high count for the tundra, when translated into a Hopla Index, would be 18 000!

Boreal mosquitoes (and some other Diptera, especially blackflies) can become extremely dense and can interfere with human activities. One must note, however, that actual interference is to a considerable degree psychological. Some humans can tolerate mosquitoes and blackflies, some cannot.

Much has been written on the effects mosquitoes have on such animals as caribou, muskox and moose. In actual fact, these animals tolerate mosquitoes and show little reaction other than ear-flapping, head-tossing or, in caribou, movement to more exposed and wind-swept sites. Many observers have confused mosquito-reactions in caribou with those caused by *Oedomagena* (the warble flies) and *Cephanomyia* (the nose-bot flies). These two genera are exceedingly rare in comparison to mosquitoes yet the presence of only one or two individuals will cause an entire band of caribou to have violent, sometimes panicky, reactions.

DOWNES (1964) thought that the two most important features of boreal environments affecting insects are (1) decreased energy input, (2) the great fluctuations in photoperiod.

DOWNES (1965) stated that an outstanding feature of Arctic insects is their tolerance of very low temperatures in winter. Survival of freezing is by means of high glycerol content or by dehydration. Once frozen, the insect is quite safe from the effects of further cooling. Such an ability to withstand freezing is called 'cold hardiness'. On the other hand, 'cold adaptation' is the ability to maintain metabolism at a higher rate at a given temperature than related species can. DOWNES (1964) reported that certain chironomids on northern Ellesmere Island could fly even at 3.5°C.

Wind is as characteristic of tundra regions as low summer temperatures. On islands constant winds result in many insects having reduced wings, or even no wings. A similar evolutionary change has occurred in some kinds of tundra insects. DOWNES (1962, 1964) reported that some species of Arctic blackflies (Simuliidae) have lost the mating swarm-flights; they mate on the ground near where they emerge. Eight of the nine true Arctic species of blackflies in Canada have gone even farther by giving up the necessity for feeding flights. The females of these species have non-functional mouthparts; their eggs develop from nutrients ingested during the earlier aquatic larval stages.

KIRCHNER (1973) measured supercooling points of a number of central European spiders. Their supercooling points are well within the range of temperatures found in the pukak space in the taiga where spiders have been found active (PRUITT, 1973) and NÄSMARK (1964) found a variety of invertebrates active.

## 6.2   Amphibians and reptiles

Among the terrestrial vertebrates the amphibians and reptiles are ectotherms. Thus these animals reach their northern limits of distribution in regular progression as one goes northward down the energy gradient (BLEAKNEY, 1958). Although the northern limits of distribution of various species of amphibians and reptiles are known with a fair degree of accuracy, we know very little of the actual factors that govern the northern limits of these animals. SPELLERBERG (1973) reviewed some the explanations which have been postulated.

Northern regions have short seasons in which these animals can be active. Shortness of active season affects the total amount of growth and fat accumulation possible. The short growing season, moreover, may be cool and may have cold snaps early and late. A basic concomitant of northern environments is permafrost and annual freezing. The distribution of the European Viper (*Viperus berus*) in Finland is controlled by soil temperatures (VIITANEN, 1967).

The winter is the critical season for northern amphibians and reptiles. These animals overwinter either subterrestrially or subaquatically. Those that overwinter below ground either must dig below the frost line or must be able to endure freezing temperatures. Such a species would be affected by permafrost. Those species that overwinter subaquatically require a body of permanent water that does not freeze and that does not 'winterkill' because of lack of oxygen.

## 6.3   Birds and mammals

In boreal endotherms temperature of the body is usually well above that of the environment but is no higher than in their relatives in more southern regions. Boreal endotherms are the birds and mammals. As in most other kinds of life there is a noticeable reduction in number of kinds between southern and northern regions. Because mammals are relatively sedentary, an analysis of their distribution might reveal insights into the effects of boreal environments on life in general. SIMPSON (1964) showed the dramatic decline in kinds of terrestrial mammals from south to north in North and Central America:

|  | Species per quadrat of 39 km² |
| --- | --- |
| Costa Rica (10°N. Lat.) | 163 |
| Mexico | 90–150 |
| South-central California | 115 |
| Arizona | 100 |
| Most of the taiga region | 45 |
| Treeline | 35 |
| Low Arctic tundra | 30 |
| Continental High Arctic tundra (70°N. Lat.) | 13 |
| Queen Elizabeth Islands High Arctic tundra | 8 |

(Simpson's analysis encompassed the continent only; the datum for the Queen Elizabeth Islands High Arctic tundra results from my own observations on Devon Island.) This apparently simple progression is complicated by vegetational zonation and post-glacial history.

Why are there more species in the tropics than in the temperate zone, more in the temperate zone than in the subarctic and Arctic? Many explanations have been offered but, in truth, none are satisfactory. Simpson restated the problem as: since each species occupies a niche, then (1) there are more niches in the tropics or (2) niches in the tropics are more fully occupied or (3) there is multiple occupation or overlap of niches in the tropics. FISCHER (1961) explained the phenomenon as a result of long-continued stable conditions with high energy gain in southern regions. Such explanations imply that there simply has not been enough time since glacial retreat for animals and plants to invade parts of the north. However, PORSILD (1951), SAVILE (1956, 1961) and DOWNES (1964) have considered the problem for Ellef Ringnes Island, probably the most rigorous low-altitude Arctic environment. They concluded that there are about as many species of animals and plants on the island as the climate allows and that the relatively recent exposure of the island to colonization cannot be invoked as an explanation for the few kinds of life present.

HAGMEIER and STULTS' (1964) analysis centred about determinations of 'Index of Faunal Change' (IFC), the percentage of species whose range limits occurred within quadrats of 6500 km². Thus, they showed that the transcontinental taiga is a real entity contrasted with the forest-tundra and Low Arctic tundra which are regions of change (Figs. 6–1 and 6–2).

Hagmeier and Stùlts' classic paper shows us some other characteristics of boreal mammals. For example, in spite of its deprecation by physiologists (e.g. SCHOLANDER, et al. 1950; IRVING, 1972) Bergmann's Rule actually describes, in general terms, a natural phenomenon. In the popular mind (and even some scientific literature) dormancy is a common way mammals meet the rigours of a northern winter. In fact, this is not so since the Eskimoan province has the lowest percentage of dormancy and the Hudsonian the third lowest rate. Dormancy seems to be an adaptation to seasonal drought rather than to seasonal cold.

## 6.4    Schools of thought on boreal endotherm adaptation

There are two schools of thought influencing concepts about adaptations of endotherms to Arctic or tundra conditions. HOFFMANN (1974) summarized the prevalent North American ideas:

1  Tundra vertebrates are of fairly recent origin.

2  Recognition that tundra small mammals have reproductive adaptations to a short growing season but no emphasis or, indeed, even mention of the stress of moulting or the juxtaposition of lactation and moulting as an adaptation. There is, however, recognition of moulting in birds as stressful.

3  Heavy emphasis on the role of thermoregulation ('cold adaptation') as a major route of adaptation to tundra conditions.

4  Great dependence on the results of field research done on the Arctic Coastal Plain of Alaska.

The research done at the Institute of Biology at Sverdlovsk exemplifies the Soviet school of thought on adaptations to tundra conditions.

SHVARTS (1963) after extensive field and laboratory analysis, classified the mammal fauna of the Eurasian Low Arctic tundra into several categories, based on distribution, habitats and characteristics of method of adaptation.

1  *Natives*  This group, limited in its distribution mainly or exclusively to the tundra or forest-tundra, consists of a group of circumpolar species

**Fig. 6–1** ESSA-8 Satellite photograph taken above central Canada 28 March 1969. Note, from north to south, the snow-covered tundra and forest-tundra, the dark mass of the taiga speckled with snow-covered lakes, the greyish tones of aspen parkland (particularly to the west of Lake Winnipeg), the snow-covered steppes. Compare this photograph with the same area represented by Fig. 6–2. (Photograph by NASA/NOAA USA ESSA-8, acquired and reproduced by courtesy of Atmospheric Environment Service, Toronto.)

clearly distinguishable in their morphology from related forms of lower latitudes. The group includes *Lemmus* and *Dicrostonyx* (lemmings), *Alopex* (white fox), *Rangifer* (caribou) and *Ovibos* (muskox), all of which are sharply differentiated on the generic or subfamily level from their relatives. Their species, however, are only weakly differentiated. Some of the species are represented over vast regions by one or a very few closely-allied subspecies. Shvarts stated that the storage of energy reserves in various forms is the most important adaptation of mammals to the tundra.

All of the native species begin to reproduce very early in the year, at times when temperature or photoperiod change cannot be invoked as causal stimuli. Lemmings can remain sexually potent throughout practically the entire year; photoperiod appears to have minimal effect on their sexual activity. The 'native' species have a lower fecundity than the tundra populations of 'cosmopolitans'. High fecundity is an advantage only in favourable conditions; when conditions are less than optimal it can become harmful.

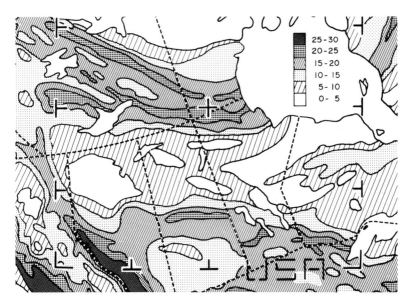

| | |
|---|---|
| ▓ | 25-30 |
| ▒ | 20-25 |
| ▨ | 15-20 |
| ░ | 10-15 |
| ⧄ | 5-10 |
| ☐ | 0-5 |

Fig. 6-2 Central Canada showing regions with equal Index of Faunal Change, according to HAGMEIER and STULTS, 1964: 128. Reference marks are the same as in Fig. 6-1. Note the low IFC's in the taiga, with relatively high IFC's in the forest-tundra and the aspen parkland, thus demonstrating the validity of the taiga as a biological entity. (Map courtesy of Dr E. M. Hagmeier, published with the permission of the Society of Systematic Zoology.)

Shvarts observed that 'natives' undergo the process of reproduction with comparatively less stress than other forms, as measured by enlargement of the adrenal glands in the females. Thus, in *Lepus* (hare), *Clethrionomys* and *Microtus* (voles) and *Sorex* (shrew) the adrenals enlarge during reproduction but in *Lemmus* and *Dicrostonyx* they do not. Shvarts observed that female lemmings retain, even during reproduction, a comparatively high content of Vitamin A in their livers. Shvarts concluded that the most important peculiarity of the population dynamics of tundra mammals is their capacity to increase their numbers markedly during a minimal span of time.

The commonly-held popular notion that tundra mammals have exceptionally effective fur is not true. With a few exceptions, such as *Alopex* and *Ovibos*, the insulating properties of the fur of the 'natives' are no better than those of the taiga populations or species. While good fur is a useful adaptation it is not one of the biological prerequisites for successful life in tundra regions.

Large tundra mammals possess a great ability to withstand the direct effects of low temperatures and some show special adaptations. Smaller tundra mammals cannot possess such perfect physical thermoregulation.

In most, the critical temperature does not differ substantially from that in tropical species. In *Lemmus obensis* and *Microtus middendorffi*, however, the critical temperature is lower than in the case of related species of more southern origin. But since the average summer temperature in the tundra is relatively very low, one must logically conclude that the lowering of the critical temperature in these two species is an adaptation, not only to winter, but also to summer conditions.

In small mammals reproduction stops at low temperatures. In lemmings, weakened individuals are very susceptible to temperatures of even 10°C. Thus microclimate and efficient habitat selection (to *avoid* the low temperatures) are extremely important to tundra small mammals.

Low temperatures are not the only conditions leading to additional energy expenditures. The period of reproduction is one such condition. In the tundra, with its short summer, the reproductive period is compressed and intense, leading to energy stress. In most mammals moulting and reproduction are usually separated in time, but in numerous tundra forms pregnancy and lactation coincide with moulting. In a number of tundra mammals migrations are common and also require energy expenditure.

Since the level of metabolism in the 'natives' is not higher than in related forms, but is actually lower, we can conclude that an important adaptation to tundra conditions is the more economic expenditure of energy.

Shvarts commented that it is significant that not only are there no tundra forms with kidneys larger than in southern populations but in *Lemmus obensis* and *Microtus middendorffi* the kidneys are markedly smaller than in southern forms of the same size (cf. LIVCHAK, 1959). In contrast, these two 'natives' possess larger hearts while the tundra populations of widely-distributed species possess hearts no larger than their southern populations.

Shvarts noted that ecizing the tundra has been accomplished in some species without substantial changes in metabolism. Those species best adapted to tundra conditions have evolved the ability to maintain normal life activity while general metabolism is lowered (economy of energy expenditure), and, at the same time, by increasing the heart size are able to increase their motor ability when needed.

Shvarts concluded that the 'natives' of the tundra evolved in the North a long time ago, before the Pleistocene. In other words, the typical tundra species came into being before the origin of the tundra as a separate physiographic-geographic zone because their adaptations are to the effects of high latitudes, not to cold or lack of trees or such secondary geographic characteristics.

2   *Cosmopolitans*   This group of mammals is widely distributed in several physiographic zones, in some cases from tundra to deserts. They play important roles in the tundra ecosystem. Thus, for example, with the

exception of the white fox, all the carnivores inhabiting the tundra belong to this group.

The tundra representatives of this group possess large body size, relatively short tail, and strong growth of winter fur. But these characteristics also occur in the taiga representatives of the group. Thus one cannot state that the major adaptations of mammals to the tundra are morphological characteristics to retain heat.

A high capacity for reproduction is the main characteristic of 'cosmopolitans', enabling them to increase spectacularly and thus play dominant roles in the forest-tundra and southern tundra ecosystems. This greatly increased fertility may have evolved during a relatively short period of time because of strong selective pressures. It seems that this characteristic, in conjunction with a wide ecological valence, allows the cosmopolitan species to be dominant without substantial specific adaptations.

3   *Forest species*   Most of the species of this group are found in more southerly parts of the tundra or penetrate into typical tundra only in certain regions. Only *Alces* (moose), *Clethrionomys rutilus* (northern red-backed vole) and *Cl. gapperi* (southern red-backed vole) penetrate far into the tundra. Most of the moose on the tundra migrate in winter. Tundra populations of *Clethrionomys* are restricted in habitat to sites with relatively mild subnivean microclimates (PRUITT, 1966).

The forest forms are the newest additions to the tundra fauna and possess no adaptations specifically for the tundra; their general adaptations to the taiga, as well as their fecundity, allow penetration of the tundra.

While the 'natives' begin to reproduce long before spring, even during winter under the snow, the forest species (as well as the cosmopolitans and the steppe species) still are controlled by the onset of this complex of environmental factors. The farther north one goes, the later spring comes, until one comes to a region where a limit is reached; not enough time to raise adequate numbers of young, to moult and to lay on fat. Thus the northern limits of range of these species are governed by phenology. It is not the severe climate, nor the character of the vegetation, nor even the presence of permafrost which limits them. The basic reason is the shortening of the summer period, a phenomenon related directly to latitude.

4   *Steppe species*   These species, especially *Microtus* (subgenus *Stenocranius*) and *Citellus* (squirrel), penetrate different regions of the tundra and the range of several is extensive. Shvarts believed that one of the main requirements for a small mammal to ecize a region with a long Arctic winter and a short summer is subnivean reproduction. Other characteristics are rapid growth rate and relatively large body size.

*Citellus* is an anomalous beast, a true hibernator in the tundra. It is remarkably restricted in occurrence to sites with a thick, well-drained soil layer, with permafrost relatively far below the surface, or those with thick zabois that melt early in the summer. Almost all burrow systems seem to be well established and it could be that many have been occupied continuously for hundreds of years. In arctic Alaska there is a close coincidence of *Citellus* burrow systems with ancient Inuit house-ruins. This may be an artefact, since the pre-contact Inuit perhaps built their houses in accord with the same factors that influence *Citellus* burrow locations.

As a whole, the group of species that originated in the steppes has no such sharply-expressed adaptations to tundra life as has the group of 'natives'. Their adaptations leave no doubt, however, that they are not a new element in the tundra fauna; their age is probably comparable to that of the tundra itself. *Microtus gregalis* (singing vole) has existed in the tundra for a very long time, certainly sufficient for speciation to occur. In the vast *gregalis*. Perhaps this is because their adaptation to the tundra environment went the peculiar routes of changes in phenology of reproduction and of a more perfect utilization of microclimates, adaptations that did not require any changes in physiology or morphology.

5   *Mountain species*   These species have disjunct distributions and upon closer examination it can be seen that all mountain species encountered in high latitudes continue to keep to mountain ranges in the tundra region. This is clearly contradictory to what has been observed for plants and perhaps other groups or animals. The vegetational and animal resemblance of the tundra to the alpine zone is usually explained by the similarity of their climates, the effects of glaciation on distribution and survival of alpine species and, finally, on post-glacial dispersal of pre-adapted species. This hypothesis is well established and buttressed by numerous observations. It is, however, quite untrue in relation to tundra mammals. Not a single mountain species that occurs in the geographic regions of the tundra has come down from the mountains to the plains, let alone become a typical tundra mammal. This applies to *Ovis* (sheep), *Marmota* (marmot), *Ochotona* (pika) and some of the Eurasian species of *Clethrionomys*.

Mountain species do not enter the typical tundra ecosystem but form their own alpine ecosystem. The adaptations of mammals to high mountains involve different mechanisms from adaptations to high latitudes. The invasion of the tundra zone by mammals is not related, then, simply to climate and adaptations to one aspect of climate (namely, low temperature) but is much more complex.

It is clear from this comparison of the North American and Soviet schools that North American boreal ecology suffers from a dearth of studies closely integrating laboratory and field observations. It also is evident that over-dependence on the Arctic Coastal Plain of Alaska for field studies has produced distorted ideas of the nature of tundra, since the Arctic Coastal Plain is a relatively small, isolated area quite atypical of most of the world's tundra.

It is noteworthy that when MACPHERSON (1965) analysed the mammalian fauna of the Canadian tundra, on the basis of Wisconsin refuges and post-Wisconsin dispersal, he arrived at categories of species remarkably similar to Shvarts'. Future detailed studies of adaptive strategies should exploit Macpherson's analysis.

# 7 Ecosystems and Foodwebs (Foodchains)

The greatest advance in biology since Darwin has been the insight that the earth's biosphere is a great engine, an engine run by energy from the sun and consisting of thousands of *kinds* of parts (each kind called a species). Energy flows through this engine by means of metabolic processes.

No machine is perfect or 100% efficient. Just so with the biosphere engine system (or 'ecosystem'). At each transfer of energy and materials there is inefficiency or loss. The end result is the well-known Eltonian pyramid in which each trophic level is considerably smaller than the one on which it feeds. The size of the first trophic level of the Eltonian pyramid is a function of the quantity of available energy. Thus tropical ecosystems have a larger first level than do taiga or tundra ones. The size of the consumer levels, however, is a function of several other factors—total energy available as food, pattern of availability of energy throughout the year, proximity of the environment to the pessimum for available fauna and, very importantly, sheer size of the fauna itself (FISCHER, 1961).

Fewer species means fewer opportunities for consumer species to evolve specializations for feeding on each other. Specializations in northern species, at least those specializations readily apparent, seem to be those concerned with the physical rather than with the biotic environment.

*Taiga productivity* GRODZINSKI (1971a and 1971b) calculated that the annual above-ground primary production of the taiga in interior Alaska (Fig. 5–2) was 2145 kg/ha, or a production of $4.23 \times 10^{10}$ J/ha/year, which is the potential manufacturer of the small mammal food supply for a cycle-average of 1086 g/ha ($6.76 \times 10^6$ J/ha) of small mammals. Grodzinski calculated that the small mammals (i.e. *Sorex cinereus* (masked shrew), *Clethrionomys rutilus, Microtus oeconomus, Glaucomys sabrinus* (northern flying squirrel) and *Tamiasciurus hudsonicus*) actually had available $5.48 \times 10^9$ J/ha of food. He calculated the annual production, assimilation and consumption of these animals. Because of the magnitude of their population cycles, production varied from $2.093 \times 10^6$ to $3.35 \times 10^4$ J/ha/year, assimilation from $1.59 \times 10^8$ to $1.67 \times 10^8$ J/ha/year, and consumption from $1.97 \times 10^8$ to $2.09 \times 10^9$ J/ha/year. Thus, because of the cycle, the consumption of food by small mammals of the taiga varies from 3 to 38% of the potential supply (GRODZINSKI, 1971b). This is a much greater proportion than that consumed in temperate zone forests but still far too small to account for population cycles.

TELFER and SCOTTER (1975) analysed productivity of the aspen parkland in the southern transition of the taiga in Elk Island National Park in Alberta. They calculated that standing crop biomass of 57.08 kg/ha of *Bison, Cervus* (wapiti) and *Alces* gave a sustained yield of 11.4 kg/ha/year.

*Tundra productivity* In temperate regions many plants can gain one-third of their total weight in one week, or several thousand kg/ha annually. In contrast, it was found by WARREN-WILSON (1964) that *Salix arctica* (Arctic willow) plants in the High Arctic of Cornwallis Island produced 30 kg/ha/year.

A summary by BLISS (1975) of studies by various investigators in the High Arctic of north-eastern Devon Island, NWT, revealed that annual above-ground primary net productivity varied from 70 kg/ha on the Polar Desert plateau to 1502 kg/ha on the relatively lush sedge-moss wet meadows. This is the equivalent of $1.34 \times 10^9$ J/ha on the Polar Desert plateau and $2.77 \times 10^{10}$ J/ha on the wet meadows. *Dicrostonyx* standing crop on beach ridges was $6.60 \times 10^5$ J/ha; *Ovibos* standing crop on sedge-moss meadows was $3.78 \times 10^7$ J/ha. The Devon Island International Biological Programme studies have shown that the main energy and materials pathways in tundra are from plants to soil organisms, thus 'detritus system' is a good term. The system acts to store materials (especially nitrogen and phosphorus) and energy in forms that plants and animals find difficult to use. Thus the low productivity may be due as much to lack of suitable gene pools as to low levels of materials and energy themselves.

OSMOLOVSKAYA (1948) diagrammed the tundra foodweb on Yamal Peninsula. Her work was especially important because she described, and quantified to some extent, the variations in energy flow caused by cyclic fluctuations in small mammal populations (Fig. 7–1). In contrast to a 'high mouse year' a 'low mouse year' resulted in snowy owls (*Nyctea scandiaca*) and white fox (*Alopex lagopus*) deserting the region entirely. Osmolovskaya outlined in some detail the food habits of the two species of jaegers (*Stercorarius longicaudus* and *S. parasiticus*). She showed that even such a specialist as the peregrine (*Falco peregrinus*) varied its diet in different years and in slightly different regions, i.e. about 100 km apart (Fig. 7–2).

Osmolovskaya's work, more than any other, clarifies and illustrates a basic characteristic of boreal foodwebs, namely, the flexibility in food habits of northern animals, i.e. boreal feeding niches are broad and flexible and boreal animals usually do not show the extreme feeding specializations characteristic of temperate and especially tropical foodwebs. This study also illustrated another characteristic of simple foodwebs—their fragility or the ease by which they are upset. When a foodweb is made up of a few species, that is, with the energy and materials undergoing few transformations and with few or no alternative routes,

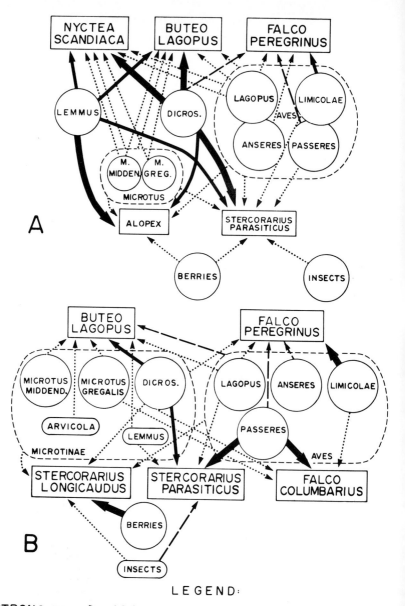

**Fig. 7–1** Shifts in a tundra foodweb. See text for discussion. (**A**) Food pathways in a year of dense vole and lemming populations. (**B**) Food pathways in a year of sparse vole and lemming populations. Yamal Peninsula, USSR. (Redrawn, with author's permission, from OSMOLOVSKAYA, 1948.)

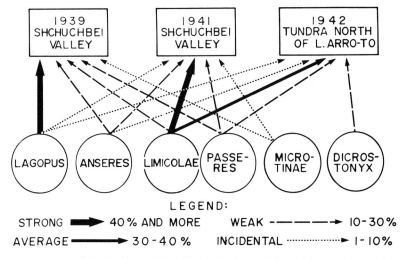

**Fig. 7-2** Shifts in feeding habits of a relatively-specialized avian carnivore, the peregrine (*Falco peregrinus*), in time and space. See text for complete discussion. Yamal Peninsula, USSR. (redrawn, with author's permission, from OSMOLOVSKAYA, 1948.)

elimination or drastic change in one of the routes causes dramatic change in all other routes. We should note that the feeding shifts observed by Osmolovskaya were on the biologically complex (and more typical) tundra of Yamal Peninsula, while MAHER (1970), on the very simple Arctic Coastal Plain of Alaska, found that jaegers there did not shift their feeding habits but continued to depend upon *Lemmus* for 90 to 100% of their food.

The fact that boreal foodwebs (tundra ones in particular) undergo such dramatic shifts regularly has led some ecologists mistakenly to insist they are not fragile but are tough and can recover. The error in this view lies in the interpretation of 'recovery'. If one gene pool were to be eliminated or *permanently* changed in size then the size, selection pressures and possible evolutionary changes of the others would be affected because there are few or no alternative sources of food. Since man's own survival in boreal regions depends on skilful management and rational use of native plants and animals, it is in his own enlightened self-interest to recognize the fragility of these foodwebs and to protect them from modification.

In a study of the ecology of a part of north-western Alaska a similar situation was pointed out (PRUITT, 1966) (Fig. 7-3). Here the terrestrial vertebrate foodweb rested on three 'legs' of primary consumers of vegetation—the microtine rodents, the Arctic ground squirrels and the caribou. All the higher consumer levels, whether jaegers, or wolves or man, depended on these three legs. The microtine rodents underwent a

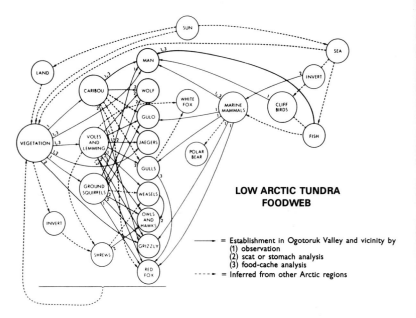

Fig. 7-3  Low Arctic tundra foodweb (PRUITT, 1966). Note how, on land, all the vertebrate carnivores depend on three main herbivores—the caribou, the voles and lemmings and the ground squirrels. Ogotoruk Valley, north-western Alaska.

violent fluctuation during the three years of study, the caribou also underwent violent fluctuations, but on a seasonal basis as they came and went on their migrations. It became clear that any outside activity that changed any one of the three 'legs' would have immediate and severe repercussions on all the other vertebrates, including man.

As an example of a naturally-occurring repetitive change, we may cite the effects of high microtine populations on *Eriophorum* (cotton sedge). TIKHOMIROV (1959) noted that on Taimyr Peninsula *Eriophorum* required more than one growing season to produce a flowering shoot. During a high population of *Lemmus* their subnivean feeding can eliminate all over-wintering partly-matured flowering shoots of *Eriophorum*. As a result vast areas show no cotton sedge flowers the following summer. A similar relationship and effect was observed on *Eriophorum* in north-western Alaska, except there the microtine was *Microtus oeconomus* (PRUITT, 1966).

The chain of relationships can be carried further. LENT (1966) observed that gravid caribou, on their migration to the fawning ground, were markedly diverted by budding *Eriophorum* freshly-exposed by melting snow. (Such shoots are notably high in protein.) It requires no great stretch of the imagination to visualize how a *Microtus* or *Lemmus* 'high' could affect and direct caribou movements.

# 8 Human Utilization of Boreal Regions

## 8.1 Pre-contact boreal man

Virtually nothing is known of the early history of man in the boreal regions. True, there are archaeological sites that enable us to distinguish successive cultures such as Dorset or Thule, but these mostly pertain to restricted, ecologically-rich coastal regions. For the vast region of the central Canadian tundra, for example, essentially nothing is known of the pre-contact human inhabitants, and the transcontinental taiga is little better represented. However, from $^{14}$C dating we know that humans have been in northern North America for at least 27 000 years, even before the maximum of the last glaciation, and were in Beringia during its isolation (IRVING and HARINGTON, 1973; HARINGTON, BONNICHSEN and MORLAN, 1975).

Apparently pre-contact boreal man was primarily Mongolian in origin. The exceptions were in Fenno-Scandia and European Russia. Undoubtedly the reasons were basically zoogeographical, although COON, GARN and BIRDSELL (1950) advanced some hotly-disputed morphological and physiological justifications.

Virtually all ideas about early boreal man result from extrapolation of conditions at the times various recent groups first contacted white Europeans. We assume cultural adaptations were distributed as they were later—basically a tundra culture and a taiga/forest-tundra culture. There were, of course, variants. The Thule culture, for example, specialized in preying on marine mammals.

Early man must have been distributed thinly in taiga and tundra. Using data given by BURCH (1972) it can be estimated that the human biomass, at the time of contact with European whites varied from 0.007 kg/ha to a maximum of 0.13 kg/ha in the forest-tundra and tundra. Boreal man was governed by a seasonal progression of harvests; caribou, moose, seals, char and whitefish—each in their season. Those regions that offered a complete sequence of food sources throughout the cycle of the seasons supported the greatest human biomass.

In only a few coastal places, where both the marine and terrestrial ecosystems could be tapped for materials and energy, did humans congregate. Elsewhere they lived in families or small groups, only coming together for short periods of time. Indeed, they could not afford to congregate, because early boreal man was a top carnivore. Over most of the boreal regions man was a predator on *Rangifer*, and also on *Alces*. MECH (1970) has observed that wolves (*Canis lupus*) cannot exceed a biomass of one wolf per 11 000 kg of prey without reaching the limit of sustained yield. Since early man was presumably in relative equilibrium with his food supply he too could not harvest more than sustained yield.

A few human cultures evolved reindeer herding. Archaeology does not record the beginning of herding with any degree of assurance. Considering the near-universal experience of caribou biologists with the tendency of new-born caribou fawns to imprint on humans, domestication must have been easy. There is no reason why it could not have occurred independently several times. While reindeer herding probably allowed an increase in human biomass undoubtedly its greatest contribution to boreal man was stability of food and materials and greatly increased mobility. On the other hand, reindeer herding locked the human group into a single repetitive routine of activities since the humans became symbiotic with the reindeer.

## 8.2   Contact and exploitation

Western man invaded the taiga and tundra both easterly and westerly from Europe. In the sixteenth century Russian settlers pushed ever eastwards along the shores of the White and Barents seas. The Cossack Semyon Dezhnev continued the exploration and in 1648 sailed through what is now called Bering Strait (SEMENOV, 1963). In Czarist times the taiga and tundra were looked upon as storehouses, inexhaustible storehouses, to be plundered continually for taxes and tributes. The native peoples were considered nothing more than machines to produce the taxes and tributes. Since the Revolution much of the northern efforts of the Soviet people have been directed towards scientific investigation of their environment and rational use of it (MOWAT, 1970). A basic theme of Soviet use of the North has been to use all of the resources of each settlement region, thus making them as independent as practicable (LAVRISHCHEV, 1969). This philosophy has had far-reaching implications for aspects as varied as politization of the indigenous human populations of the regions to shipment of fresh vegetables, from the sex and age classes of the invading humans to expansion of the reindeer industry (ARMSTRONG, 1965, 1966). It has gone far to prevent the worst abuses of the profit-oriented single-use exploitation of the North which is still characteristic of Canada and Alaska. None the less, there still occur in northern USSR examples of single-use exploitation with disastrous ecological effects (LAVROV, 1974).

Greenland is usually considered to be an example of good colonial administration (DUNBAR, 1947b). The small home country, Denmark, has actively fostered ecological research for many years. Thus, when a climatic warming trend degraded seal hunting, enough basic information was available to enable the people to turn to cod fishing for their livelihood (HOBART and BRANT, 1966). Foreign exploitation of Greenland's resources has been limited. In spite of some lapses the Danish administration has resulted in maintenance of relatively stable ecosystems and healthy people with intense pride in being, not Danish, not Inuit, but Greenlanders. Such pride is fostered by a literature printed in Greenlandic and regular radio broadcasts in Greenlandic.

The Canadian North was invaded some 400 years ago, first by whalers who decimated the Atlantic right whales, then the walrus, the bowhead whales and the blue whales. The fur traders took over next. It took them somewhat longer to decimate the beaver, the wapiti, the grizzly bear and the muskox. The fur traders, in concert with the churches and the Royal Canadian Mounted Police all acted to reduce the native peoples of the taiga and tundra to distorted shades of temperate-zone Caucasians in environments where temperate-zone activities were ecologically futile (LANTIS, 1957, 1966; JENNESS, 1966; HOBART, 1968). Unrestrained mineral prospecting resulted in a shocking increase in wildfire in the spruce-lichen caribou winter range. The resulting collapse of the caribou forced the people off the land and into slum conurbations.

The Canadian North, particularly the Arctic Archipelago (Fig. 1–4), suffered greatly from the activities of the 'explorers', especially the Franklin Search and the contest for the Pole. These parties, in order to feed themselves, killed large numbers of caribou, muskox, polar bears and walrus. One explorer, the American R. E. Peary, even partially financed his expeditions from the sale of muskox robes and walrus ivory. It is quite possible that the muskox of eastern Ellesmere Island and north-western Greenland have yet to recover from such ruthless exploitation.

Ecological stability in the Canadian North is made more difficult because of two factors: (1) the political division of responsibility for resources between Provincial, Territorial and Federal governments and (2) the official policy of the Federal Government (CANADA, 1972) that the future of the North lies in exploitation of extractive, non-renewable resources (PRUITT, 1970b; BANFIELD, 1970). This policy is actually not greatly different from the Czarist policy for Siberia whereby that region was but a treasure house to be plundered. The situation is exacerbated by policies that allow, even promote, control of resources by non-Canadian companies (CUTLER, 1975).

The seventeenth-century Czarist expansion eastwards across Siberia extended beyond Bering Strait to the Commander Islands and Alaska, where the independent entrepreneurs or 'promyshlenniki' invaded the Aleutian Islands, south-western and south-eastern Alaska, extirpated Steller's Sea Cow (*Hydrodamalis*) and decimated the populations of fur seals (*Callorhinus*) and sea otter (*Enhydra*). The Czarist government in Alaska was followed, after a time, by the United States' military government which supervised the transfer from one set of exploiters to another. After the decline of the easily-exploited marine mammal resources in southern Alaska emphasis shifted northward, to the bowhead whales of the north-west and north coast, and the fur of the interior of the subcontinent. Commercial whaling lasted only about 50 years; the whales were saved by the lucky perfection of the electric light, the motor car and a change in ladies' fashions (all of which collapsed the markets for whale oil and baleen). Fur trapping has fluctuated widely from all-out assaults by both natives and non-natives during times of general economic depression to only light harvests by dedicated 'bush Indians' when boom times have drawn others to city jobs. Like the Canadian government, the United States government has also had a predeliction for emphasis on extractive, non-renewable resources. During the gold rush and subsequent decades the human population of Alaska increased far beyond the ability of the land, or the existing supply lines, to support it. Commercial market hunting of wild mammals and birds became an important source of protein for the invaders, with the consequent collapse of some species populations. Unrestrained prospecting and mining activities led to a massive increase in wildfire.

BUCKLEY (1957) and FOOTE and WILLIAMSON (1966) showed that renewable resources of animal origin were the mainstay of Alaskan existence, during a period when their protection and management was minimal and when the Territorial, and later, State government was controlled by mining interests. The oil discoveries at Prudhoe Bay, and the associated 'energy crisis', have not helped the cause of ecosystem stability in Alaska (or the Canadian North, either) (KLEIN, 1970; CALDWELL, 1972).

Recently the resurgence of native awareness and the activities of such organizations as Inupiat Paitot (Alaska), National Indian Brotherhood (Canada) and the Inuit Tapirisat of Canada are strengthening the resistance of northern peoples against further exploitation (CROWE, 1974). They have not gained enough power as yet to stop the worst of the white man's exploiting schemes—the trans-Alaska pipeline, the Mackenzie Valley pipeline, the James Bay power project in Quebec and the diversion of the Churchill River in Manitoba (GILL and COOKE, 1974; CASNP, 1974).

Northern vegetation, particularly the taiga, is especially susceptible to fire. In very recent times there has arisen the idea that 'fire is good for the forest'. The concept stems from the influence on professional foresters by the lumber and paper cartels, because their interests lie in the earlier stages of forest succession. In some southern forest types such 'management' may perhaps be a useful tool, but in boreal regions, particularly the taiga, fire is unequivocably disastrous. It is disastrous in terms of permafrost degradation (VIERECK, 1973a, 1973b), soil (LUTZ, 1952), timber production (LUTZ, 1956), fur production and caribou and reindeer production (SCOTTER, 1964).

## 8.3   Conclusion

It has been discussed earlier how the boreal regions are desperately poor in basic solar energy (Fig. 1–2). The North cannot have more than a small, scattered human population that is permanent and living on the land. The caribou is the best factory known for capturing the scant solar energy and turning it into a form that man can utilize. In their present depleted state the caribou are scarcely a usable resource, but if they were allowed to recover to even half their primeval numbers they could furnish some 6 000 000 kg of meat and fat annually (KELSALL, 1968; PRUITT, 1970b). No matter what he does, man's activities set vegetation succession back to early or pioneer stages; as long as man remains in the North there will be a sufficiency of areas in early stages of succession. Thus a major effort should always be the protection of mature vegetation from disruption, particularly by fire.

The savannahs, and the taiga, lack large native domestic animals. The recent success of 'game ranching' schemes in Africa suggests we might examine the concept for the taiga (TELFER and SCOTTER, 1975). Indications are that there is a higher return in meat per unit area from properly managed and harvested wild caribou than from domestic reindeer. The recent Soviet success in domesticating moose points to the possibility of completely revising our concepts of human use of the taiga, indeed of all the boreal regions (KNORRE, 1959, 1961, 1969; SABLINA, 1973). Since domestic animals are the basis of all civilization, research on domestic *Alces* and *Rangifer* should be high on the priority list of all boreal governments.

In regions of reduced energy input, permafrost, slow vegetational succession and low mammalian biomass, all human schemes must be examined critically from an ecological perspective. The decision to implement any scheme must come only after we know it will not degrade the ecosystem. The ecological history of the white man's invasion of boreal regions teaches us one unassailable fact: any obligatory restriction to profit-oriented free enterprise as the sole economic system allowed is incompatible with ecosystem stability. Under this system no scheme or action dare be undertaken unless a profit is predicted and, conversely, any scheme or action that promises a short-term profit is automatically considered desirable (STONEMAN, 1972). The boreal regions are too important to the future of mankind to allow their use to be governed by only one preconceived economic system. The only valid criteria for determining use are ecological ones.

# 9  Research Methods and Procedures Peculiar to Boreal Regions

Ever since the days of belief in Ultima Thule and the Northwest Passage to Cathay, boreal regions have suffered from the attentions of purveyors of false or misleading information. Even relatively modern scientific literature has had its share of myths or traditions about the 'Mysterious North'. The unique combination of physical and biological peculiarities that go to make up 'boreal conditions' requires the scientist and student be alert to sometimes obvious, but more often subtle, pitfalls when designing field experiments or outlining observations to be taken.

Because of great differences in productivity of various vegetation types an area may, in one visitor's short experience, be interpreted as showing a high lemming population, when in truth it may be a low. This is a common error and it leads to what should be a general research directive: the state of boreal small mammal populations can be told only by sequential, standardized sampling carried out over the complete cycle or fluctuation. The *cycle* is the basic terrestrial ecological unit in boreal regions (Figs. 7–1 and 7–2). Any study touching on any phase of terrestrial ecology in the North must allow for the cycle. It is central in determining harvest quotas of large ungulates, furbearers and even fruits and berries.

In the past thirty years a vast amount of time and effort has been spent in studying chemical 'cold adaptation' in boreal animals. The impetus resulted from a preoccupation by fund-granting agencies (particularly military ones) with an almost-mystical belief in the possibility of discovering some physiological mechanism that, if understood and controlled, would allow temperate-zone humans to live in the North with only southern clothing, equipment and attitudes. It should be evident by now that the mechanisms by which mammals, at least, have adapted to the North are primarily in the realms of behaviour, population structure and ability to overcome stresses of reproduction, moulting, migration or photoperiod changes. Adaptation to boreal conditions is a far more complicated business than mere ability to withstand cold. Research on 'cold adaptation' is legitimate in its own right, but our understanding of boreal ecology would be much more advanced if we were not sidetracked into looking on it as the strategy whereby birds and mammals ecize boreal regions.

The winter snow cover continues virtually to be ignored by those who should be concerned. For example, in the classic monograph on North American snowshoes the author stated, 'It is important to note that the type of snow is not a barrier' (DAVIDSON, 1937). On analysis of the situation it was found (PRUITT, 1970c) that there were a number of sharply-expressed relationships between various construction details in snowshoe types and the major snow regions of Canada.

KREBS (1964) similarly dismissed weather phenomena as important controls in *Lemmus* and *Dicrostonyx* cyclic fluctuations. It remained for FULLER (1967) to rework Kreb's published data and to derive a perfectly logical and reasonable explanation for the population changes, an explanation that, by application of Occam's Razor, is probably closer to the truth than Kreb's more involved explanation.

A major key to understanding cyclic fluctuations of populations of boreal small mammals lies in analysis of autumnal and vernal critical periods and the winter snow cover (Fig. 4–1) (PRUITT, 1957; FULLER, 1967; KUKSOV, 1969). This is one of the directions in which future studies of population fluctuations of boreal animals should go (HARE, 1971; SIIVONEN, 1956; FULLER, STEBBINS and DYKE, 1969).

Another possible factor governing population fluctuations in boreal small mammals is cyclic ecosytem productivity. Observations implicate some mechanisms(s) whereby productivity waxes and wanes in a regular, recurrent fashion (PRUITT, 1968, 1972; MYRBERGET, 1973). The actual increase or decrease of any one species may be governed by whether the species was in an ecological stage to react to the productivity changes.

The great Russian naturalist A. N. FORMOZOV (1946) classified animals on the basis of their ecological reactions to snow: *Chionophobes*—species that do not inhabit snowy regions, that avoid snow, and include the small cats, steppe antelope, steppe sand grouse, black partridge, many small terrestrial birds. *Chioneuphores*—species that can withstand winters with considerable snow, and include moose, reindeer, wolverine, wolf, fox, many voles, moles and shrews. *Chionophiles*—species whose ranges lie completely or almost completely in regions of hard and continuous winters. These species have characteristic adaptations (winter-white colouration, winter peculiarities of foot-coverings, etc.) which undoubtedly were perfected by snow cover taking part in selection. They include ptarmigan, varying hare, Arctic fox, collared lemming, caribou, etc.

In regions with only occasional api the snow that collects on trees (qali) is primarily of transitory aesthetic interest. Qali forms, ideally, under cold, windless conditions (Fig. 5–2). A superficially-similar but quite different phenomenon is *kanik*, which forms when relatively warm, moist (or supercooled) air passes over cold objects, such as vegetation, wires or towers. Maximum accumulation of kanik occurs at temperatures only slightly below freezing and with slight to moderate winds (ZAMORSKII, 1955; MILLER, 1964). Sometimes qali may accumulate upon already-formed kanik, confusing the analysis.

In the windless taiga of central Alaska qali assumes ecological importance (PRUITT, 1958). It is one of the agents initiating forest succession. If a spruce departs from the vertical it is doomed to be broken, some day, because it will accumulate qali. When a tree breaks, adjacent trees become susceptible to qali-breakage and the glade grows until it is sufficiently large that wind circulation prevents massive accumulation of qali. In the glade the broken spruces die and the rain of dead needles chokes out the feather-mosses on the forest floor. Thus seeds have a good site for germination. Deciduous trees invade—alders, birches, aspens and willows. These trees mature and die and in their leaf litter young spruce can germinate. They mature and the spruce forest eventually is restored at the site, to await further qali-breakage.

Qali is also of direct economic importance to man's activities. Power lines frequently are snapped either by qali-broken trees or by a heavy accumulation of qali on the wires themselves. Some boreal rural electrification cooperatives have met the challenge by periodically flying helicopters at a low altitude along the most vulnerable lines so that the rotor blast cleans the qali from the power wires.

Spruce stands straight and tall and resists the qali; alders and birches are limber and bend with it and recover in the spring. When the trees are bent over by qali their tender growing tips are brought within the range of the hares (*Lepus americanus* or *L. timidus*) which feed extensively on them. Snow-caves form under the bent trees and in very cold weather, when even the hares avoid the infinite heat sink of the night sky, these caves are radiation-refuges for them.

Thus in the taiga qali has a variety of ecological effects. Qali accumulation is exceptionally difficult to measure quantitatively. A 'qalimeter' enables objective comparisons of qali accumulation, both geographically and chronologically, to be made (PRUITT, 1958, 1973).

It has been shown how important api analysis is to the understanding of the ecology of small mammals; the role of api in the lives of large mammals is equally influential. In the central Canadian taiga the areas of suitable api wherein caribou overwinter are surrounded by 'fences' of unsuitable api (PRUITT, 1959a). In years when the thickness, hardness and density gradients are well-developed they restrict the animals' movements. If the areas of suitable snow conditions coincide with areas of good food supply the caribou can pass the snow season in good condition. If, however, the areas of suitable snow conditions coincide only with areas of poor food supplies (e.g. burns) then the caribou may survive only with difficulty.

This reasoning may be extended into time and the situation during a period of climatic warming can be considered. As the temperate-zone snow conditions intrude poleward into the taiga, the caribou are squeezed between their advance and the treeless tundra where the winds still rework and harden the snow cover. The optimum overwintering environment for caribou seems to be subarctic taiga with relatively thin api that is not modified by near-thawing atmospheric conditions or by wind. Thus, changing nival conditions is just as reasonable an explanation for the extirpation of such groups as *Rangifer tarandus dawsoni* on the Queen Charlotte Islands (BANFIELD, 1961) and *R. t. eogroenlandicus* in East Greenland (DEGERBØL, 1957) as are more abstract explanations based on genetic drift. Both these now-extinct subspecies lived in fringe regions where slight climatic shifts could result in worsening nival environments.

If the taiga snow cover has so many ecological implications consider the hard, wind-worked tundra snow cover or upsik (Fig. 3–1). Above the upsik is the moving snow or siqoq (Figs. 1–3 and 1–4) which is either consolidated into a succession of drift forms or moves along and above the snow surface (ORLOV, 1961). The siqoq becomes stabilized for varying periods of time and forms drifts.

On a flat, relatively unobstructed surface many of the siqoq particles advance in groups. A group assumes a characteristic arrowhead shape with the point upwind, a gradually sloping upwind face and a lee slope which is abrupt and concave laterally, thickest at the tang of the arrowhead. These drifts are known popularly as barkhans but more accurately in Inuit as *kalutoganiq*. Kalutoganiq migrate downwind as the particles are exposed on the windward face, are moved over the surface of the drift, and then temporarily are immobilized on the steep lee slope. Whenever the wind slackens the kalutoganiq become consolidated through the processes of sublimation and re-crystallization.

Later winds may erode away the kalutoganiq, producing sculptured forms which have great beauty but which are exceedingly difficult to traverse. The sculpturings are known as zastrugi (Russian) or skavler (Norwegian) but more accurately in Inuit as *kaioglaq*. Zastrugi or skavler refer to surface sculpturings in general; kaioglaq refers to large, hard sculpturings while the word *tumarinyiq* (also Inuit) refers to small zastrugi or 'ripple marks'.

Kaioglaq eventually may be eroded away completely and the particles regrouped downwind again into kalutoganiq. A late stage of kaioglaq is the formation of overhanging drifts or *mapsuk*. The windward point of a ridge of kaioglaq is eroded faster at base level than above it, thus forming the characteristic anvil tip which points upwind.

Upsik may completely fill valleys of small streams (Fig. 1–3) and form thick drifts below the crests of hills, on the downwind side. Such drifts may not melt until late the following summer and are called *zaboi* (Russian). They retard plant growth and, in those spots where they do not melt until late in the summer, their presence may prevent certain species from living. In extreme cases they may prevent all plants from growing on the site where they form (BILLINGS and BLISS, 1959). These bare spots are then subject to intense cryopedological processes which may lead eventually to characteristic tundra landscape forms. Zaboi may also be regulators of mesic habitats in an expanse of otherwise rather xeric conditions. The characteristic plants associated with zaboi indicate places where a highway or railroad should not be run, if there is concern about it being blocked in winter (GJAERVOLL, 1956).

In some spots on the tundra where upsik completely fills a small stream valley eddy currents may scour out the snow and produce a tremendous cavity. These scoured spots are known in Inuit as *aŋmaŋa*. In the folk-knowledge of northern peoples aŋmaŋa around large rocks are known as good places to set traps or snares since here the subnivean vegetation or soil is exposed. SULKAVA (1964) showed how aŋmaŋa around field hay sheds (Fig. 5–1) are important in the ecology of partridge (*Perdix perdix*) and hare (*Lepus europaeus*) in Finland. In some tundra regions white fox hunt from one aŋmaŋa to another, since ptarmigan may be found feeding or gravelling there.

The Inuit seem to have concentrated their folk-study of upsik on the surface configurations. It would appear that they distinguish only one kind of basal layer to upsik or api, namely *pukak*. In the official meteorological lexicon pukak goes by the inelegant (and actually incorrect) name of 'depth hoar'.

In nature several kinds of pukak are present. The Lapps differentiate these (ERIKSSON, 1976) since the variations are important to reindeer which obtain their food from under the pukak layer (Fig. 3–1).

Because of the effect of wind, tundra upsik is remarkably constant year to year. Such predictability allows a numerical expression of its ecological effect to be arrived at. Such effects for a particular tundra region as they correlated with the fluctuating populations of small mammals there were calculated (PRUITT, 1966, 1970a) and a 'Snow Index' (SI) derived:

$$SI = C(\Sigma TD)$$

where $C$ = snow cover of a plot expressed as a percentage of the area of the plot covered by it,
$T$ = thickness (in cm) and
$D$ = density of each discrete layer of snow as measured when exposed in a vertical profile.

The Snow Index correlated nicely with numbers, species and fluctuations of the populations of small mammals in the fauna. With more refined observations time could perhaps be included as a term in the SI formula because, as has been shown, the details of the onset, duration and disappearance of the upsik are also powerful ecological factors.

The present SI formula was derived after consideration of the relative effect the factors had on small mammals in the Low Arctic tundra. In other boreal situations different formulae may be necessary; other organisms also will require different formulae. The Snow Index concept is in its infancy and much further research is needed.

A Snow Index for each species of vertebrate animal in a region would be a valuable tool in wildlife management. For example, whitetail deer (*Odocoileus virginianus*) have well-defined thresholds to thickness, hardness and density of api. The northern edge of their range shifts as these parameters of the api fluctuate. In such regions traditional temperate-zone wildlife management techniques are of little value and may even be harmful. A Snow Index for whitetail deer, along with a collection of proper environmental data, would enable the nivally substandard range to be delimited so that money formerly spent there on management might be spent elsewhere with more effect. Knowledge of the nival ecology of a species is especially pertinent to introductions of exotics.

An environmental factor as all-pervasive and powerful as snow cover must have been as important an ecological factor in the past as it is in the present. Indeed, several phenomena of the geologically-recent past are understandable when interpreted as snow phenomena.

Continental glaciation itself is but an imbalance between snow accumulation and disapperance—more falls each winter than melts the following summer. Thus the 'centres' of glaciation are not necessarily small; they may be very large, an appreciable part of a continent. The time of change between glacial and inter-glacial is probably also a time of environmental oscillation before the climate settles down to being pure inter-glacial or pure glacial. As far as animals are concerned, the periods of climatic change (that is, pre-glacial or post-glacial) may be times of severe ecological stress as the winters vary markedly between little snow and excess snow. Thus nival conditions are a perfectly reasonable explanation for the dramatic shifts in range and the extinctions which occurred in post-glacial time. For example, snow cover is as logical an explanation for the post-Wisconsin extinction of so many large mammals in North America as are some other more involved and tortuous explanations (e.g. overhunting by primitive man; see MARTIN and WRIGHT, 1967; MARTIN, 1973). The nival explanation is even more logical when we realize that virtually all the species extirpated were (if we can extrapolate from their most similar surviving relatives) chionophobes.

A knowledge of snow cover and boreal conditions enables man to understand better the lives of his own ancestors. Thus BURCH (1972) clarified the ecological role of early *Rangifer* hunters by extrapolation from his wide knowledge of present and recently-past caribou-hunting Indians and Inuit, as well as a synthesis of recent knowledge of caribou ecology. He showed that many of the anthropologists' best-loved attributes of early human cultures are quite incompatible with *Rangifer* hunting as a way of life.

Many popular works picture early man striding resolutely across Beringia, up to his bare knees in soft api and wearing nothing but a fur loin cloth with a wolf skin draped around his bare shoulders. Our best indications are that Beringia was tundra-like and had a very cold winter climate with little snow. The Inuit of the region, even in these days of synthetic fabrics and modern technology, when on extended hunting trips in winter, use finely-tailored caribou skin clothing. Thus their Beringian ancestors undoubtedly already possessed well-made skin clothing, walked and hunted on bare ground or exposed vegetation and over rock-hard upsik.

There is a widespread legend about the boreal regions in post-Beringian times, namely that the large mammals existing then were decimated suddenly by intense cold. The story has affected non-biologists' interpretations of boreal ecology (e.g. HERBERT, 1971). It has arisen, perhaps, from the fact that some of the carcasses found in permanently-frozen ground are in remarkably good states of preservation. Remarkably good, that is, for being 10 000 years old! Stomach contents are identifiable, for example. But the carcasses were not 'quick-frozen'. One chapter of the legend had mammoth steaks served at a banquet of the Academy of Sciences! In actual fact, only the sled dogs of the excavation expedition would touch it (TIKHOMIROV, 1961; FARRAND, 1961). When similar carcasses are sluiced out of frozen ground by the high-energy water jets at the gold placer mines in Alaska they are definitely not fit to eat; indeed, the stench is overpowering. Environmental conditions today, in fact, are such that similar large mammals could be preserved in cold, quaking bogs or thermokarst pits. There is no need to conjure up a sudden, catastrophic chilling of the climate.

The over-dependence on temperature readings, when the concern should really be with heat capacity and heat flow, has already been discussed. This can scarcely be over-emphasized in boreal ecology. Air that loses heat becomes denser, sinks and flows downhill. Thus topographic concavities become puddles of cold air and the channels leading to the concavities may be sites of extreme windchill. Paradoxically, the cold in such channels and concavities is more pronounced in calm than in windy weather. In windy regions air temperatures from a meteorological station a few kilometres distant may be extrapolated, without great inaccuracy. In calm regions, especially if there are hills or valleys, this should never be done. In such regions air temperature differences of as much as 20 Celsius degrees are common between the top of a hill 100 metres above the valley and the valley floor itself.

The importance of infrared in heat flow in boreal regions must not be forgotten. Undoubtedly the most important member in infrared flux in the North is the infinite heat sink of the night sky. This heat sink can affect any animal exposed to it, as well as the api surface itself. The thin layer of extremely cold air at the api surface affects boreal ecology in two ways: (1) general chilling of the region by radiation exchange and (2) negative effects on small mammals and snow insects which may venture onto the surface. The basic outline of this phenomenon has been sketched, but clearly the appreciation of its importance needs greater dissemination.

If there is one ecological factor which, more than any other, characterizes boreal regions, periodicity must be nominated—diel, seasonal and cyclical. This periodicity dominates all aspects of boreal ecology and must be taken into account in all northern biological research. Until the various cycles of boreal regions are understood the various animals and plants concerned cannot truly be 'managed'.

Although understanding of the behaviour of moisture in boreal regions is actually remarkably good, it has failed to penetrate higher levels of government and industry. Prefabricated houses, with insufficient insulation and lacking vapour barriers, are still shipped north; boots and clothing made of closed-cell insulation are still sold in northern stores as the latest innovation. Every northern school room should have a framed motto 'Get rid of the moisture but keep your insulation dry.' Popular appreciation of the role of atmospheric moisture would be advanced immeasurably if there existed a cheap, easy way to measure it at low temperatures so that every thermometer had an easily-readable hygrometer mounted alongside it.

Almost everywhere one stands in the North frozen ground is not far beneath one's feet. Permafrost exists only because of a delicate balance among incoming energy, outgoing energy, soil, vegetation, present and past climates and other factors (see section 3.3). The great integrating stratum in the balance is the vegetation, the insulator. Almost all of mans' present activities disrupt vegetation—vehicle passage, forest fires, tundra fires, land clearing for agriculture and buildings, even concentrated foot traffic—and thus tend to degrade permafrost.

In northern activities it has long been axiomatic to protect permafrost at all costs. The most successful method has been to elevate structures on pilings. No really successful surface method has been perfected. Thus sub-structure insulation (i.e. gravel pads) eventually fails because all insulation can do is slow down heat flow, not eliminate it. In the case of linear structures such as pipelines, even piling is unsatisfactory since the activities of the pile-driving and pile-supplying machinery lead to permafrost degradation. Clearly, much technological progress is sorely needed in this field.

A phenomenon peculiar to boreal regions is 'ice fog', which forms only at ambient temperatures below about −40°C. Any source of atmospheric moisture is also a potential source of ice fog—open water, human breath, chimneys, internal combustion engines, fires, etc. The ice particles in ice fog are microscopic but do not form a true aerosol since they aggregate and slowly settle out. None the less, ice fog can become more and more dense and the layer over the source thickens and enlarges with time.

Ice fog is a severe problem in northern cities—air travel stops, vehicle traffic slows to a crawl and the concentration of air pollutants increases dramatically. Indeed, ice fog is the greatest deterrant to northern urbanization and urban industrialization. Research on the problem should be high on the priority list for boreal human ecologists (ROBINSON, THUMAN and WIGGINS, 1957; BENSON, 1969, 1970).

A central factor in man's use of any environment is vegetational productivity. It has been shown that boreal regions have fewer species and lower productivity than more southern regions. Moreover, boreal vegetation has a higher proportion of lichens than does southern vegetation. Thus the vegetation base is more suceptible to all kinds of air-borne pollutants than that of non-boreal regions. Contamination with radioactive particles is perhaps the best-studied (PRUITT, 1963) but others such as $SO_2$ (GILBERT, 1965; SCHOFIELD and HAMILTON, 1969; ERICKSON, 1973) and fluorides (Le BLANC, COMEAU and RAO, 1971) may be equally dangerous.

Boreal ecological research is difficult, in terms of personal comfort, safety, expenses and complexity of individual adaptations in clothing, transport and activities, attitudes and psychology. The sheer distances involved in the North mean that supply lines for boreal field research parties and stations may be long and expensive. Aircraft charter and freight charges frequently make up major items in budgets for northern research. In some instances fund-granting agencies have rejected worthwhile proposals for field research because they did not appreciate the necessity of the high costs involved. These deterrants mean that boreal ecologists are relatively few and, perforce, dedicated.

All field ecologists tend to be attracted to one region more than others and to become identified with it. Boreal ecologists are more susceptible than most to this condition, not only because of the challenge of northern research and the scientific rewards but because of the very real lure of the Northland. No other scientific discipline can offer such extra attractions as encounters with 'The Throng' of thousands of migrating caribou that vibrate the very earth with their footfalls, or the glimpse of a great bull moose trotting through the golden-leaved aspens on a bright autumn day, or the skin-prickling sound of a wolf's song from a moonlight-flooded taiga hillside.

> The summer—no sweeter was ever;
> The sunshiny woods all athrill;
> The grayling aleap in the river,
> The bighorn asleep on the hill.
> The strong life that never knows harness;
> The wilds where the caribou call;
> The freshness, the freedom, the farness—
> O God! How I'm stuck on it all.

> The winter! The brightness that blinds you,
> The white land locked tight as a drum,
> The cold fear that follows and finds you,
> The silence that bludgeons you dumb.

The snows that are older than history,
The woods where the weird shadows slant;
The stillness, the moonlight, the mystery,
I've bade 'em goodby—but I can't.

(From 'The Spell of the Yukon', by Robert Service. Reprinted by permission of McGraw-Hill Ryerson, Limited.)

# References

The source literature of Boreal Ecology is scattered in numerous, sometimes obscure, journals and books, not only in English but in Russian, Danish, Norwegian, Swedish and Finnish. The starred (*) articles are those I consider of major importance.

AHTI, T. (1961). *Arch. Soc. 'Vanamo', Painettu*, **10**, VII, 3 pp.

*ALEXANDROVA, V. D. (1970). *IUCN Publ.*, New Ser., No. 16, 93–114.

ALLEE, W. C., EMERSON, A. E., PARK, O., PARK, T. and SCHMIDT, K. P. (1949). *Principles of Animal Ecology*. Saunders, Philadelphia, 837 pp.

ANNERSTEN, L. G. (1964). Investigations of permafrost in the vicinity of Knob Lake 1961–2. *Montreal, McGill Univ. Subarctic Research Paper*, No. 18, 51–137.

ARMSTRONG, T. E. (1965). *Russian Settlement in the North*. Cambridge University Press, Cambridge.

ARMSTRONG, T. E. (1966). In: *The Arctic Frontier* (ed. R. St. J. Macdonald). Univ. Toronto Press, Toronto, 57–88.

BALANTSEVA, I. A. (ed.). (1960). *Map of Vegetation Types of the USSR, 1:10 000 000*. Moscow, Ministry of Geology and Protection of the Land, USSR, 1 sheet 72 × 96 cm.

BANFIELD, A. W. F. (1954). *J. Mammal.*, **35**, 104–7.

BANFIELD, A. W. F. (1961). *Nat. Mus. Can. Bull.*, 177, *Biol. Ser.*, No. 66, 137 pp.

BANFIELD, A. W. F. (1970). *IUCN Publ.*, New Ser., No. 16, 207–9.

BENSON, C. S. (1969). *Polar Record*, **14**, 783–90.

*BENSON, C. S. (1970). *Cold Regions Research Engineering Laboratory, Research Rept.*, No. 121.

BILLINGS, W. D. and BLISS, L. C. (1959). *Ecology*, **40**, 388–97.

BILLINGS, W. D. and MORRIS, R. J. (1951). *Amer. J. Botany*, **38**, 327–31.

BLEAKNEY, J. S. (1958). *Nat. Mus. Can. Bull.*, 155, *Biol. Ser.*, No. 54, 119 pp.

BLISS, L. C. (1975). Devon Island, Canada. In: Structure and Function of Tundra Ecosystems (eds. T. Roswall and O. W. Heal). *Ecol. Bull.*, **20**, 17. Swedish Natural Science Research Council, Stockholm.

*BLÜTHGEN, J. (1970). *Ecol. and Conservation*, **1**, 11–33.

BRITTON, M. E. (1957). *Proc. 18th. Biol. Colloq. Oregon State College*, 26–61.

BROWN, R. J. E. (1960). *Arctic*, **13**, 163–77.

BROWN, R. J. E. (1966a). *Proc. Intl. Permafrost Conf. (1966)*, 241–6.

BROWN, R. J. E. (1966b). *Proc. Intl. Permafrost Conf. (1966)*, 20–5.

*BROWN, R. J. E. (1970). *Permafrost in Canada, its Influence on Northern Development*. Univ. Toronto Press, Toronto, 234 pp.

*BRYSON, R. A. (1966). *Geog. Bull.*, **8**, 228–69.

BRYSON, R. A., IRVING, W. N. and LARSEN, J. A. (1965). *Science*, **147**, 46–8.

BRYSON, R. A., and WENDLAND, W. M. (1967). In: *Life, Land and Water* (ed. W. J. Mayer-Oakes). Univ. Manitoba Press, Winnipeg, 271–98.

BUCKLEY, J. L. (1957). *Biol. Papers Univ. Alaska*, No. 1 (rev.), 33 pp.

BUDYKO, M. I. (1963). Table 6 in: FLOHN, H. (1969). *Weather and Climate*. McGraw-Hill, New York and Toronto, 253 pp.

*BURCH, E. S. Jr. (1972). *Amer. Antiquity*, 37, 339–68.

*CALDWELL, M. (1972). In: *Socialism and the Environment*, Bertrand Russell Peace Foundation Ltd., Nottingham, 25–58.

CANADA (1972). *Ottawa, DIAND Publ.*, QS-3015-000-HB-A-1, 40 pp.

CASNP (1974). (Canadian Association in Support of Native Peoples.) Bull., 15, 22 pp.

CHURCH, N. S. (1960). *Jour. Exper. Biol.*, 37, 186–212.

*CHURCHILL, E. D. and HANSON, H. C. (1958). *Bot. Rev.*, 24, 127–91.

COON, C. S., GARN, S. M. and BIRDSELL, J. B. (1950). *Races*. Thomas Publishers, Springfield (Illinois).

CORBET, P. S. (1972). *Acta Arctica*, 18, 43 pp.

COURT, A. (1948). *Bull. Amer. Meterol. Soc.*, 29, 487–93.

CRAIG, B. G. and FYLES, J. G. (1960). *Ottawa, Geol. Surv. Can., Paper* 60–10, 21 pp.

*CROWE, K. J. (1974). *A History of the Original Peoples of Northern Canada*. McGill-Queen's Univ. Press, Montreal.

CUTLER, M. (1975). *Can. Geog. J.*, 90, 4–19 (et. seq.).

DAVIDSON, D. S. (1937). Snowshoes. *Mem. Amer. Philos. Soc.*, 6, 207 pp.

*DEEVEY, E. S. (1958). *Sci. Amer.*, 199, 114–22.

DEGERBØL, M. (1957). *Acta Arctica*, 10, 57 pp.

DOLGIN, I. M. (1970). *Ecol. and Conservation*, 1, 41–61.

DOWNES, J. A. (1962). *Can. Entomol.*, 94, 143–62.

DOWNES, J. A. (1964). *Can. Entomol.*, 96, 279–307.

DOWNES, J. A. (1965). *Ann. Rev. Entomol.*, 10, 257–74.

DREW, J. V. and TEDROW, J. C. F. (1962). *Arctic*, 15, 109–16.

*DRURY, W. H. Jr. (1956). *Contrib. Gray Herbarium, Harvard University*, No. CLXXVIII, 130 pp.

*DUNBAR, M. J. (1947a). *Canadian Field-Nat.*, 61, 12–14.

DUNBAR, M. J. (1947b). *Commerce Journ.* (March), 69–109.

ELSNER, R. W. and PRUITT, W. O. Jr. (1959). *Arctic*, 12, 20–7.

ERICKSON, D. L. (1973). *Alternatives*, 2, 27–31.

ERIKSSON, O. (1976). *Meddelanden från Växtbiologiska institutionen, Uppsala*, 2.

EVERNDEN, L. N. and FULLER, W. A. (1972). *Can. J. Zool.*, 50, 1023–32.

FARRAND, W. R. (1961). *Science*, 133, 729–35.

FISCHER, A. G. (1961). *Amer. Scientist*, 49, 50–74.

FLOROV, D. N. (1952). *Zool. Zhurn.*, 31, 875–82.

FOOTE, D. C. and WILLIAMSON, H. A. (1966). In: *Environment of the Cape Thompson Region, Alaska* (eds. N. Wilimovsky and J. N. Wolfe), USAEC-PNE-481, 1041–1107.

*FORMOZOV, A. N. (1946). *Materials for Fauna and Flora USSR, Zoology Section*, New Ser., 5, 152 pp.

FOSTER, J. B. (1961). *J. Mammal.*, 42, 181–98.

FRASER, J. K. (1959). *Arctic*, 12, 40–53.

FREUCHEN, P. and SALOMONSEN, F. (1958). *The Arctic Year*. Putnam's, New York, 438 pp.

FROHNE, W. C. (1956). *Public Health Reports*, 71, 616–21.

*FULLER, W. A. (1967). *La Terre et la Vie 1967*, 97–115.

FULLER, W. A., STEBBINS, L. L. and DYKE, G. R. (1969). *Arctic*, 22, 34–55.

*GATES, D. M. (1962). *Energy Exchange in the Biosphere*. Harper and Row, New York, 151 pp.

*GEIGER, R. (1965). *The Climate Near the Ground*. Harvard Univ. Press, Cambridge, 611 pp.

GIDDINGS, J. L. (1961). *Univ. of Alaska, Studies of Northern Peoples*, No. 1, 159 pp.

GILBERT, O. L. (1965). *Brit. Ecol. Soc. Sympos.*, No. 5, 35–47.

GILL, D. (1975). *Can. J. Earth Sci.*, **12**, 263–72.

GILL, D. and COOKE, A. D. (1974). *Polar Record*, **17**, 109–27.

GJAERVOLL, O. (1950). *Svensk Bot. Tidskr.*, **44**, 387–440.

GJAERVOLL, O. (1956). *Norske Videnskabers Selskab, Skrifter*, No. 1, 405 pp.

GORHAM, E. (1957). *Quart. Rev. Biol.*, **32**, 145–66.

*GRIGOR'EV, A. A. (1946). *The Subarctic*. Academy of Science Press, Institute of Geography, Moscow–Leningrad, 171 pp.

*GRODZINSKI, W. (1971a). *Acta Theriol.*, **16**, 231–75.

GRODZINSKI, W. (1971b). *Ann. Zool. Fennici*, **8**, 133–6.

HAGMEIER, E. M. and STULTS, C. D. (1964). *Syst. Zool.*, **13**, 125–55.

HAMILTON, N. J. and HEPPNER, F. (1967). *Science*, **155**, 196–7.

*HAMMEL, H. T. (1956). *J. Mammal.*, **37**, 375–8.

HARDY, J. D. and STOLL, A. M. (1954). *J. Applied Physiol.*, **7**, 200–11.

*HARE, F. K. (1970). *Trans. Royal Soc. Can.* (Ser. 4), **8**, 393–400.

HARE, F. K. (1971). *Rep. Kevo Subarctic Res. Sta.*, **8**, 31–40.

HARE, F. K. and RITCHIE, J. C. (1972). *Geog. Rev.*, **62**, 333–65.

HARINGTON, C. R., BONNICHSEN, R., and MORLAN, R. E. (1975). *Can. Geog. J.*, **91**, 42–8.

HERBERT, W. (1971). *Polar Deserts*. Collins, London, 128 pp.

*HOBART, C. W. (1968). *J. Amer. Indian Educ.*, **7**, 7–17.

HOBART, C. W. and BRANT, C. S. (1966). *Can. Rev. Socio. and Anthrop.*, **3**, 47–66.

HOCKING, B. and SHARPLIN, C. D. (1965). *Nature*, **206**, 215.

HOFFMANN, R. S. (1958). *Ecology*, **39**, 540–1.

HOFFMANN, R. S. (1974). Chapter 9 in: *Arctic and Alpine Environments* (eds. J. D. Ives and R. G. Barry). Methuen, London.

*HOK, J. R. (1969). *A Reconnaissance of Tractor Trails and Related Phenomena on the North Slope of Alaska*. U. S. Dept. Interior, Bureau of Land Mgt., 66 pp.

HOPKINS, D. M. and SIGAFOOS, R. S. (1951), *U.S. Geol. Surv. Bull.*, 974–6, 51–101.

HOPLA, C. E. (1964–5). *Brooklyn Entom. Soc. Bull.*, 59/60, 88–127.

HUSTICH, I. (1953). *Arctic*, **6**, 149–62.

HUSTICH, I. (1966). *Reports Kevo Subarctic Res. Sta.*, **3**, 7–48.

IRVING, L. (1972). *Arctic Life of Birds and Mammals Including Man*. Springer-Verlag, Berlin, 192 pp.

IRVING, W. N. and HARINGTON, C. R. (1973). *Science*, **179**, 335–40.

*JENNESS, D. (1966). In: *The Arctic Frontier* (ed. R. St. J. Macdonald). Univ. Toronto Press, Toronto, 120–9.

KARPLUS, N. (1952). *Ecology*, **33**, 129–34.

KELSALL, J. P. (1968). *The migratory barren-ground caribou of Canada. Ottawa, Can. Wildl. Serv. Monogr.*, No. 3, 340 pp.

KEVAN, P. G. (1970). 'High Arctic insect-flower relations: the interrelationships of arthropods and flowers at Lake Hazen, Ellesmere Island, NWT, Canada'. Upubl. Ph.D. Thesis, Univ. of Alberta.

KEVAN, P. G. (1972). *Can. Field-Nat.*, **86**, 41–4.

*KEVAN, P. G. and SHORTHOUSE, J. D. (1970). *Arctic*, **23**, 268–79.

KHLEBNIKOV, A. I. (1970). *Zool. Zhurn.*, **75**, 801–2.

KIL'DYUSHEVSKII, I. D. (1956). *Bot. Zhurn.*, **41**, 1641–6.

KIRCHNER, W. (1973). In: *Effects of temperature on ectothermic organisms* (ed. W. Wieser). Springer-Verlag, Berlin, 271–9.

KLEIN, D. R. (1970). *IUCN Publ.*, New Ser., No. 16, 209–42.

KNORRE, E. P. (1959). Pechoro-Ilych National Park, *Trudy*, **7**, 5–122.

KNORRE, E. P. (1961). Pechoro-Ilych National Park, *Trudy*, **9**, 5–113.

KNORRE, E. P. (1969). Moscow Society of Naturalists, *Trudy*, **35**, 13–19.

KOSHKINA, T. V. and KHALANSKI, A. S. (1962). *Zool. Zhurn.*, **41**, 604–15.

KREBS, C. J. (1964). *Arctic Inst. North Amer. Tech. Pap.*, No. 16, 104 pp.

KROG, J. (1955). *Physiologia Plantarum*, **8**, 836–9.

KUKSOV, V. A. (1969). Scientific Research Institute of Agriculture in the Far North (Institute of Polar Agriculture), *Trudy*, **17**, 176–9.

LANTIS, M. (1957). *Oregon State College, Biol. Colloq.*, **18**, 119–30.

LANTIS, M. (1966). In: *The Arctic Frontier* (ed. R. St. J. Macdonald). Univ. Toronto Press, Toronto, 89–119.

LAVRISHCHEV, A. (1969). *Economic Geography of the USSR.* 'Progress' Publishers, Moscow, 379 pp.

LAVROV, N. P. (1974). In: *Acclimatization of Game and Commercially-Utilized Mammals and Birds in the USSR (Part II)* (eds. M. P. Pavlov, I. B. Korsakov and N. P. Lavrov). Volgo-Vyatsk Book Press, Kirov, 356–63.

Le BLANC, F., COMEAU, G. and RAO, D. N. (1971). *Can. J. Bot.*, **49**, 1691–8.

LENT, P. C. (1966). *U.S. Atomic Energy Commission, PNE*—481, 481–517.

LIVCHAK, G. B. (1959). Acad. Sci. USSR. Ural Branch, Salekhard Station, *Trudy*, **1**, 280–92.

LUTZ, H. J. (1952). *Proc. Alaska Sci. Conf. 1952*, 54–5.

*LUTZ, H. J. (1956). *U.S. Dept. Agric. Tech. Bull.*, No. 1133, 121 pp.

LYMAN, C. P. (1943). *Bull. Mus. Comp. Zool. Harvard Univ.*, **93**, 393–461.

*MACPHERSON, A. H. (1965). *Syst. Zool.*, **14**, 153–73.

MAHER, W. J. (1970). *Wilson Bull.*, **82**, 130–57.

MARTIN, P. S. (1973). *Science*, **179**, 969–74.

MARTIN, P. S. and WRIGHT, H. E. (eds.) (1967). *Pleistocene Extinctions*. Yale Univ. Press, New Haven, 453 pp.

MASON, B. J. (1961). *Sci. Amer.*, **204**, 120–31.

*MECH, L. D. (1970). *The Wolf: the Ecology and Behavior of an Endangered Species.* Natural History Press, New York, 384 pp.

MILLER, D. H. (1964). *U.S. Forest Service Research Paper*, PSW-18.

MOSQUIN, T. (1966). In: *The Evolution of Canada's Flora* (eds. R. L. Taylor and R. A. Ludwig). Univ. of Toronto Press, Toronto, 43–65.

*MOWAT, F. (1970). *Sibir.* McClelland and Stewart, Toronto, 313 pp.

MYRBERGET, S. (1973). *Oikos*, **22**, 220–4.

MYSTERUD, I. (1966). *Fauna*, **19**, 79–83.

NÄSMARK, O. (1964). *Zool. Revy*, **1**, 5–13.

NOVIKOV, B. G. and BLAGODATSKAYA, G. I. (1948). *Acad. Sci. USSR. Doklady*, LXI, 577–80.

ØRITSLAND, N. A. (1970). In: *Antarctic Ecology, Vol. 1* (ed. M. W. Holdgate). Academic Press, New York, 446–70.

ORLOV, N. I. (1961). In: *Role of Snow Cover in Natural Processes* (ed. M. I. Iveronova). Moscow, Acad. Sci., USSR, 258–64.

ORVIG, S. (ed.). (1970). Climates of the Polar Regions, Vol. 14. In: *World Survey of Climatology* (ed. in chief H. E. Landsberg). Elsevier Publishing Co., London.

*OSMOLOVSKAYA, V. I. (1948). Academy of Science USSR. Institute of Geography, *Trudy*, **41**, 5–77.

PEIPONEN, V. A. (1970). *Ecol. and Conservation*, **1**, 281–8.

PÉWÉ, T. L. (1957). *Proc. 18th Biol. Colloq. Oregon State College*, 12–25.

PINTER, A. J. and NEGUS, N. C. (1965). *Amer. Journ. Physiol.*, **208**, 633–8.

*POLLETT, F. (1968). *Newfoundland, Dept. of Mines, Mineral Resources Report*, No. 2, 226 pp.

PORSILD, A. E. (1951). *Can. Geog. J.*, **45**, 27.

*PRUITT, W. O. Jr. (1957). *Arctic*, **10**, 130–8.

PRUITT, W. O. Jr. (1958). *Ecology*, **39**, 169–72.

PRUITT, W. O. Jr. (1959a). *Arctic*, **12**, 158–79.

PRUITT, W. O. Jr. (1959b). *J. Mammal.*, **40**, 139–43.

PRUITT, W. O. Jr. (1960). *Sci. Amer.*, **202**, 60–8.

PRUITT, W. O. Jr. (1963). *Audubon Magazine*, **65**, 284–7.

PRUITT, W. O. Jr. (1966). In: *U.S. Atomic Energy Commission*, PNE-481, 519–64.

*PRUITT, W. O. Jr. (1967). *Animals of the North*. Harper and Row, New York, 173 pp.

PRUITT, W. O. Jr. (1968). *Mammalia*, **32**, 172–91.

PRUITT, W. O. Jr. (1970a). *IUCN Publ.* New Ser., No. 16, 33–40.

*PRUITT, W. O. Jr. (1970b). *Trans. Royal Soc. Can. Ser. IV*, **8**, 373–85.

*PRUITT, W. O. Jr. (1970c). *Ecol. and Conservation*, **1**, 83–90.

PRUITT, W. O. Jr. (1972). *Aquilo. Ser. Zool.*, **13**, 40–4.

*PRUITT, W. O. Jr. (1973). *Techniques in Boreal Ecology. Part A—Environmental Analysis. Part B—Animal Populations and Activity*. 16mm colour teaching films produced by Instructional Media Centre, Univ. of Manitoba.

PRUITT, W. O. Jr. and LUCIER, C. V. (1958). *J. Mammal.*, **39**, 443–44.

RIKHTER, G. D. (1962). *Snow Cover, its Distribution and Role in the Economy of Nature.* Acad. Sci. Press, Moscow, 272 pp.

*ROBINSON, E., THUMAN, W. C. and WIGGINS, E. J. (1957). *Arctic*, **10**, 89–104.

ROWAN, W. (1938). *Biol. Rev.*, **13**, 374–402.

ROWE, J. S. (1966). In: *The Evolution of Canada's Flora* (eds. R. L. Taylor and R. A. Ludwig). Univ. of Toronto Press, Toronto, 12–27.

*SABLINA, T. B. (1973). *Domesticating Moose*. 'Science' Publishing House, for Moscow Society of Naturalists, Moscow, 82 pp.

SALMI, M. (1970). *Ecol. and Conservation*, **1**, 143–54.

SALOMONSEN, F. (1939–40). *Dansk naturhistorisk forening Videnskapelige meddelelser*, **103**, 491 pp.

SALT, R. W. (1961). *Ann. Rev. Entom.*, **6**, 55–74.

SAVILE, D. B. O. (1956). *Amer. Midl. Nat.*, **56**, 434–53.

SAVILE, D. B. O. (1961). *Nat. Mus. Can., Nat. Hist. papers*, No. 9, 6 pp.

*SAVILE, D. B. O. (1972). *Ottawa, Can. Dept. of Agriculture, Research Branch, Res. Monog.*, No. 6, 81 pp.

SCHOFIELD, E., and HAMILTON, W. L. (1969). *Papers 20th Alaska Sci. Conf.*

SCHOLANDER, P. R., HOCK, R., WALTERS, V. and IRVING, L. (1950). *Biol. Bull.*, **99**, 259–71.

SCHWABE, W. W. (1956). *Ann. Bot.*, New Ser., **20**, 587–622.

*SCOTTER, G. W. (1964). *Ottawa, Canadian Wildlife Service, Wildl. Mgt. Bull.*, Ser. 1, No. 18.

SEMENOV, Yu. N. (1963). *Siberia; its Conquest and Development*. Wiley, New York, 480 pp.

SETON, E. T. (1911). *The Arctic Prairies*. Scribner's, New York, 308 pp.

SHACKLETTE, H. F. (1963). 'Influences of the soil on boreal and arctic plant communities'. Unpubl. Ph.D. Thesis, Univ. of Michigan.

*SHVARTS, S. S. ('S. S. SWARTZ'). (1963). Academy of Science USSR, Ural Branch, Institute of Biology, *Trudy*, 33.

SIIVONEN, L. (1956). *Finnish Game Foundation, Papers on Game Research*, No. 17, 30 pp.

SIMPSON, G. G. (1964). *Syst. Zool.*, 13, 57–73.

SIPLE, P. A., and PASSEL, C. F. (1945). *Proc. Amer. Philos. Soc.*, 89, 177–99.

SPELLERBERG, I. F. (1973). In: *Effects of Temperature on Ectothermic Organisms* (ed. W. Wieser). Springer-Verlag, Berlin and New York, 239–46.

STEFANSSON, V. (1945). *Arctic Manual*, Macmillan, New York, 556 pp.

STOLL, A. M. (1954). *Federation Proc.*, 13, 146.

STONEMAN, C. (1972). In: *Socialism and the Environment*. Bertrand Russell Peace Foundation Ltd., Nottingham, 59–104.

SULKAVA, S. (1964). *Aquilo. Ser. Zool.*, 2, 17–24.

SVIHLA, A. (1956). *J. Mammal.*, 7, 378–81.

*TEDROW, J. C. F. (1970). *Ecol. and Conservation*, 1, 189–206.

TEDROW, J. C. F. and CANTLON, J. E. (1958). *Arctic*, 11, 166–79.

*TELFER, E. S., and SCOTTER, G. W. (1975). *J. Range Mgt.*, 28, 172–80.

*TIKHOMIROV, B. A. (1959). *Interrelationships of the Animal World and the Vegetation of the Tundra*. Academy of Science Press, Moscow–Leningrad, 104 pp.

TIKHOMIROV, B. A. (1961). *Proc. Univ. Durham Phil. Soc.*, Ser. A., 13, 241–8.

TIKHOMIROV, B. A. (1970). *Ecol. and Conservation*, No. 1, 35–40.

*TYRTIKOV, A. P. (1959). *Ottawa, N.R.C. Transl. No. 130, Div. of Bldg. Research*, 34 pp., mimeo.

VIERECK, L. A. (1965). *Arctic*, 18, 262–7.

VIERECK, L. A. (1970). *Ecol. and Conservation*, 1, 223–33.

VIERECK, L. A. (1973a). *Permafrost: North American Contribution to the Second International Conference*, 60–7.

*VIERECK, L. A. (1973b). *J. Quaternary Res.*, 3, 465–95.

VIITANEN, P. (1967). *Ann. Zool. Fennici*, 4, 472–546.

WARNECKE, G. (1958). *Proc. X Int. Congr. Entom.*, 1, (1956), 719–30.

WARREN-WILSON, J. (1964). *Ann. Bot.*, 28, 71–6.

*WASHBURN, A. L. (1956). *Geol. Soc. Amer. Bull.*, 67, 823–66.

WILLIAMS, G. P., and GOLD, L. W. (1958). *Engineering Institute of Can. Trans.*, 2, 91–4.

YASHINA, A. V. (1961). In: *Role of Snow Cover in Natural Processes* (ed. M. I. Iveronova). Academy of Science Press, Moscow, 137–65.

*ZAMORSKII, A. D. (1955). *Atmospheric Ice*. Academy of Science Press, Moscow–Leningrad, 377 pp.

ZHUKOV, A. B. (ed. in Chief). (1966). *Forests of the USSR*, 5 vols. Vol. 1—Forests of the northern and central taiga of the European part of the USSR. 'Nauka' Press, Moscow, 457 pp.